The Heavens Declare
An Introduction to Astronomy

Thomas Williams

Liberty University

Kendall Hunt
p u b l i s h i n g   c o m p a n y

Cover image © Shutterstock.com

www.kendallhunt.com
Send all inquiries to:
4050 Westmark Drive
Dubuque, IA 52004-1840

Published in the United States of America

CONTENTS

Getting Started—A Discussion About Science

Chapter 1

> *"Just as all things speak of God to those who know Him,*
> *and all things are revealed to those who love Him,*
> *just so they stay hidden from those people who do not seek Him and do not know Him."*

—Blaise Pascal

Welcome to Astronomy!

Astronomy is one of the oldest of all pursuits. People have been tracking the movements of the stars and planets for thousands of years. Genesis 1 introduces us right away to "the greater light" and the "lesser light," as well as a sky that is filled with stars. Like many other ancient cultures, the nation of Israel would use these celestial objects "for signs, and for seasons, and for days, and years." Unlike the other ancient cultures, however, Israel was called to worship the Creator of those objects rather than the objects themselves.

Later in the Old Testament, the Psalmist tells us that "the heavens declare the glory of God." That was about 3,000 years ago, long before anyone had any thought of telescopes and all observations were naked eye. The night sky was limited to the Moon, a few thousand stars, and five special wanderers that we know as planets. While anyone willing to go out and look at a clear night sky today would surely agree that what can be seen with the naked eye is indeed glorious, how much more glorious are the heavens with the aid

Figure 1.1

© AstroStar/Shutterstock.com

1

of modern technology? The discoveries of the past 400 years should not diminish the words of the Psalmist, but rather magnify them.

Today, astronomy is the branch of science that involves the study of the entire universe, from what we can observe naked eye, to the faintest and most distant galaxies that are barely detectable with our best telescopes. It is the study of all celestial objects and events. It involves studying the world of the incredibly small (high-energy particles, atoms, molecules), as well as the incredibly large (stars, star clusters, galaxies). It boggles our minds with enormous distances, and it captures our imaginations with such exotic objects as black holes, pulsars, and exoplanets.

In subsequent chapters, we will take a brief survey of various celestial objects, many of which we can still enjoy with unaided eyes, but which we know much more about today thanks to telescopes and other technologies. The emphasis will be on our own solar system, although we will also wander out to the stars. However, before we begin, we need to take a critical look at the state of science today and make some observations that will be important to keep in mind throughout our journey.

Follow the Science?

The global pandemic of 2020 popularized the mantra of "follow the science." What became increasingly unclear was what exactly that mantra meant. It did, however, attempt to harness the power and privilege enjoyed by the word "science" in our culture. Throughout the pandemic, cable news networks brought us a parade of "experts" whose stories changed and conflicted with those of equally respected "experts." For anyone paying attention, 2020 was a good time to question what we know about science. Basic questions about what science is and how science works became more relevant. Unfortunately, rather than getting answers to such questions, we got slogans, soundbites, and lectures. "Follow the science" became little more than virtue signaling. "Science is truth" became a defense against anyone questioning the authority of an expert. Rather than a tool of discovery, science became a weapon to wield against political rivals and unruly taxpayers. There was red science and blue science and whatever "truth" the scientists and bureaucrats may have been referring to was obscured in the resulting distrust and hysteria. In 2020, questioning the expert du jour got you labeled as anti-science—the new scarlet "A." The experts and doubters alike may have a general misunderstanding about what science is and how science works.

© FamVeld/Shutterstock.com

Figure 1.2

Confusion over what constitutes "good science" and the limits of science are no less common when exploring astronomy. For example, when NASA proclaims that "the universe is teeming with life," is that good science? Is it science at all? When Carl Sagan introduced his popular program

Cosmos decades ago with these words—"The cosmos is all that is or was or ever will be"—was that science? When astronomy textbooks confidently put forward just-so stories about the origin of the universe, the origin of the Earth, and the origin of life, is that the final word on those subjects? The underlying questions in each case are part of a topic commonly called "the nature of science," and attempts to answer those questions depend on philosophy, not science.

1.1 The Nature of Science

Before we proceed further, it must be clearly stated that science has led to many tremendous successes. What we typically call scientific research has saved and improved lives. The use of scientific methodologies delivered multiple COVID-19 vaccines in record time. Science can help us feed the world, achieve medical breakthroughs, generate new technologies, and yield a host of other incredibly valuable outcomes. Nothing in this chapter is meant to somehow demean or deny the value of science. There are concerns however, when it comes to how well we understand the nature of science, and how science can be used as a pretense for bad behavior.

That phrase, "the nature of science" deals with a variety of questions related to what science is, and how science works. These are deceptively complex questions that cannot be answered by using science. In other words, science cannot define itself. The questions of what science is and how science is expected to work are not themselves in the realm of science but are instead philosophical questions. This may be worth highlighting—*not every question is a science question*. That may sound obvious but for many in our culture it comes as a surprise.

Think of science as a tool, the "novum organum" (literally "new tool") that Francis Bacon wrote about in the early 1600s. Then consider the simple analogy of a hammer. A hammer is incredibly useful for many things—demolition, roofing, framing, finish work, etc. However, as soon as you have to cut a 2x4, shovel snow from your driveway, or get that tiny screw back into your eyeglasses, you no longer have the right tool. That does not diminish the value of having a hammer in your toolbox, but not every job can be done with a hammer.

© Ollyy/Shutterstock.com

Figure 1.3

Let us look at this another way. Suppose you need open-heart surgery. Do you choose a doctor based on his or her theological training? If your car will not start, do you search for an expert on 19th-century Impressionist art? Of course, these questions are ridiculous. However, many people, including many Christians, have accepted the notion that science is some sort of authority when it comes to questions about ethics, religion, history, or politics. When you accept the false premise that science is the only means to arrive at truth, then you are no longer dealing with science. Instead, you are dealing with scientism, which is a philosophical position, not a scientific one. Scientism represents yet another misconception about science, but it has been embraced by many. "Trust the science" is applied to every aspect of life, even when by its very nature science cannot possibly address these other areas. Like other misconceptions about science, the problem is not with science per se, but rather with the culture and its lack of an understanding about the limits of science.

One way to describe science then, is that it is a tool to use to try to gain knowledge and solve problems. It is not a person. Routinely we have heard phrases like these: "According to science…" "Science tells us…" "Science has proven…" Returning briefly to the pandemic, "science" did not establish policies or sign executive orders. Science did not censor critics, send COVID-19 patients into nursing homes, or politicize itself as if it were running for office. Because science is so frequently anthropomorphized, the lines are blurred between science and the people who are involved, whether scientists, politicians, or media. Science is something people do, and these people are no less flawed than anyone else. If it was not obvious before the pandemic, it became very clear that scientists have their own biases and agendas. They can be affected by politics, pettiness, and power. In the end, we always hope that science as a tool will lead to results that somehow transcend the confusion; but clearly science is an intensely human endeavor that can be misused and misrepresented by various stakeholders.

It is a sad situation that today many Christians feel compelled to view science as the enemy. If you take that as the message of this chapter, then you will have missed the mark entirely. Science has its roots in Christianity, as a quick review of the history of science would demonstrate. You will be hearing more about the likes of Copernicus, Kepler, and Newton, and it is important to see that history does not support the current aberration that science and faith are opposed to each other. These early scientists pointed to their faith as the motivation to do science in the first place. Equally sad though, is how many Christians mindlessly accept anything that scientists have to say as if the ultimate authority is indeed the word of the scientist.

Compounding the problem is the fact that there is no universally accepted definition of science, even among philosophers of science. Likewise, there is no single, universally accepted "scientific method," despite what most of us have been taught. Historically, we did not have the words "science" or "scientist" in common use until well into the nineteenth century; centuries after some of the most important astronomers made their contributions. However, by applying the work of philosophers of science, we can attempt to identify and discuss some common ground when it comes to describing science.

Coming Up With a Description

Whether we go back to the 1600s and review the work of Francis Bacon, or look at some of the most recent comments from the National Academy of Sciences, one hallmark of science is that it is based on "empirical data." Science is based on some combination of making observations, running tests, taking measurements, etc. This builds in some limitations, as we must rely on our senses, and we must work with current, detectable phenomenon. Our observations and experiments must be repeatable. The exact details may vary from one field of science to the next. A physicist may go about collecting data very differently than a biologist, which makes sense given the differences in what they intend to study or what problem they hope to solve. Regardless, the hope is that careful, skillful, repeated investigations and data collection will allow us to move to a second hallmark of science, namely the creation of a logical argument.

A logical argument may manifest itself in a number of ways. Depending on the nature of the question and the evidence that is collected, scientists may discover new patterns that lead to a breakthrough. They may propose a new system of classification or organization. They may add important details to an existing scientific principle, or they may propose a new discovery altogether. Perhaps most important to scientists is whether or not their argument allows them to make accurate predictions to further confirm their conclusions. "Good science" is not simply determined by how much evidence can be collected. The ability to show that your

Figure 1.4 Francis Bacon

argument can serve to predict what will happen under a variety of circumstances is crucial. Scientists like Galileo and Einstein had bold ideas and made bold predictions that could not be tested right away. Eventually those predictions were confirmed and helped to further validate their work.

If we can think of empirical data as the evidence we collect, then creating a logical argument can be thought of as how we interpret the evidence. This is where things get interesting. While everyone eventually has access to all the same evidence, not everyone reaches the same conclusions. One grave misconception about the nature of science is that by following a method ("the" scientific method) the facts should somehow speak for themselves and scientists should all be forced to agree on the outcome. Poorly written textbooks and well-intentioned teachers taught children for years that there was a mechanical series of steps—the scientific method—by which we arrived at our scientific truths. What tends to get downplayed is the role of interpretation. The same evidence can be interpreted differently. Ptolemy saw the sunrise as evidence for geocentrism. Copernicus saw the sunrise and concluded that it was consistent with heliocentrism. The same observations or evidence could be given radically different interpretations. Keep this in mind and we will return to it later.

Francis Bacon proposed what is sometimes called induction or the inductive method. At the time, this was largely a response to the failures that he saw in traditional Greek approaches which relied almost exclusively on deductive methods. With deductive reasoning, you begin with a set of broad, general assumptions that are accepted as being true, and you then apply them to specific questions. You might be familiar with this in the form of a geometric proof or syllogism. For example, if we start with the assumption that all triangles have three sides and that the sum of the angles is always 180 degrees, there are a host of questions we can answer regarding particular examples of triangles. The problem of course is that this method only works if your beginning assumptions are correct, and by

Figure 1.5

1600 there were plenty of examples that Bacon could point to where deductive reasoning failed in the sciences. His solution then (the early version of a "scientific method") was to turn the situation upside down. Rather than using general ideas to answer questions about the particulars, he would look at lots of particulars by observing, testing, and investigating to arrive at a conclusion about what the general principle was.

Consider the following from Robert Pirsig's "Zen and the Art of Motorcycle Maintenance":

> "Inductive inferences start with observations of the machine and arrive at general conclusions. For example, if the cycle goes over a bump and the engine misfires, and then goes over another bump and the engine misfires, and then goes over another bump and the engine misfires, and then goes over a long smooth stretch of road and there is no misfiring, and then goes over a fourth bump and the engine misfires again, one can logically conclude that the misfiring is caused by the bumps. That is induction: reasoning from particular experiences to general truths.
>
> Deductive inferences do the reverse. They start with general knowledge and predict a specific observation. For example, if, from reading the hierarchy of facts about the machine, the mechanic knows the horn of the cycle is powered exclusively by electricity from the battery, then he can logically infer that if the battery is dead the horn will not work. That is deduction."

You might be familiar with two scenes from the classic movie, "The Princess Bride." First, in "the battle of wits" between Westley and Vizzini, the dizzying display of deductive reasoning on the part of Vizzini leads to his downfall. His assumptions are flawed and inadequate and, perhaps most importantly, he did not anticipate the possibility of Westley poisoning both glasses of wine. In a second scene, we see the inductive method at work in "the pit of despair," where Count Rugen can run repeated tests and make observations regarding the effects of his "machine" on Westley. "I'm sure you've discovered my deep and abiding interest in pain…so I want you to be totally honest with me on how the machine makes you feel. This being our first try, I'll use the lowest setting…" In the end, Vizzini's deductive methods leave him dead, while the effects of Rugen's machine leave Westley only "mostly dead."

While scientific inquiry may still make use of some deductive elements, it was the purposeful shift to induction that in many ways was revolutionary. Bacon seemed convinced that his inductive method would serve the added purpose of making the results of any inquiry objective. The method would cancel out any potential bias on the part of the scientist. In fact, Bacon seemed to suggest that reason and background (i.e., experience/training) were not as important as long as you followed the right method. There are scientists today who like using the term "objective science." We will revisit this later and hopefully you can see that this is another misconception about science.

Setting aside claims of complete objectivity in science for now, we can frame the third hallmark of science as follows. Science must be open to skeptical review. An important attribute of science is that its discoveries and claims are always tentative and subject to revision in light of new discoveries, including the detection of errors. After scientists have run their tests, collected their evidence, and then drawn their conclusions (logical argument), their tests, evidence, and argument must be open to critique and criticism. This may take the form of self-critique, in which case the scientists keep reworking and improving and sometimes even abandoning their initial attempts as they recognize possible flaws in their own work. For at least the past century, skeptical review has generally meant peer-review to see if someone's work was worthy of publication. Other scientists scrutinize the work and decide whether or not it has merit. If Bacon thought his method of induction might itself produce objective truth, the modern version of that misconception can be found in the peer-review process.

Consider the possibility that all of your "peers" are wrong! The history of science has no shortage of examples of this. If you are a true revolutionary—perhaps a Newton or an Einstein—you are likely a step (or more) ahead of the current consensus of your peers. In the early twentieth century, Alfred Wegner proposed the idea of "continental drift." His idea was dismissed and ridiculed, in part because the consensus among geologists at the time was that the Earth was solid, so there was no way to imagine a mechanism for any movement. Once the model changed to a more dynamic Earth (core, mantle, crust) the consensus

Figure 1.6

shifted and today plate tectonics—the updated version of continental drift—is the dominant paradigm. The point is that peer-review may fail to recognize legitimate scholarship.

Conversely, the peer-review system can pass along flawed ideas and faulty research if it fits with the way some peers look at the world. There are many peer-reviewed articles which, although they appeared in the most prestigious journals of their time, promoted such horrific ideas as the eugenics movement in pre-WWII America or the supposed value of lobotomies as a tool for mental health. During the pandemic, two very well-known, respected journals published research about the use of anti-malarial drugs in the treatment of COVID-19 patients. The articles that were printed made claims that fell into step with the partisan politicization of the topic, and the political implications apparently suited the peer-reviewers. However, it immediately came to light that the research itself was a failure by any standard, and the two journals quietly retracted their support of the "science" shortly after publication. For such serious journals, this was a great embarrassment. It raised questions about how the work ever cleared peer-review in the first place, and the political aspect appeared to be the lone explanation.

Scientist and philosopher Charley Dewberry shares his own experience with the peer-review process in his book, *Saving Science*. Dewberry's research was in stream ecology and salmon recovery in the Pacific Northwest. He submitted his research to a journal, and it was reviewed by three well-known scientists, each with their own area of expertise. One "peer" recommended accepting the work as it was, without revision. The second recommended accepting the work with some minor revisions. The third basically said it had no value and he would never clear it for publication. Dewberry writes, "As a good empiricist, I was dumbfounded by the responses. How was it possible that the same paper could elicit three radically different reviews?" As Dewberry also noted, "Reviewers cannot help but evaluate the theories and beliefs of an author in the light of their own theories and beliefs." The third reviewer in this case saw Dewberry's work as unimportant based on his own personal and professional biases. Do not confuse the peer-review system with objectivity!

1.2 Two Kinds of Science

Figure 1.7 The lab is just one possible setting for operational science.

Figure 1.8 Historical science tries to explain unseen events from the past.

Given the general characteristics of science and scientific methodologies, the argument can be posited that there is a distinction to be made between science which seeks to answer questions regarding events or objects that can be repeatedly tested and observed, and events or objects from the past which cannot. As early as the 1930s there were science departments making such distinctions, to delineate differences between what were sometimes called "laboratory sciences" as opposed to something like anthropology. With laboratory science (also known as operational or observational science), we can repeatedly test, observe, and collect data in real time whether in a formal lab setting or out in the field. For the anthropologists traveling to dig sites to excavate bones and other remains, the dynamic was not the same. They were searching for clues and trying to answer questions from the unseen past, an alternate approach sometimes called historical science.

Harvard astronomy professor Owen Gingerich discusses the distinction in his book *God's Planet*. Gingerich shares the story of how Harvard sought to update their general education program, including what type(s) of science students needed. Stephen Jay Gould developed a very popular gen-ed class in evolutionary geology. The concern was raised that students were only getting one kind of science that was strongly historical in nature but weak on any experimentation. It was perceived as being an historical science rather than an operational science. The solution was to require students to "take a half course in each type of science" to try to create a balance between science that was "strongly…historical in nature" and science that was "undergirded by natural laws from which conclusions can be deduced and matched with experimental results."

In geology, the distinction is routinely made between "physical geology" and "historical geology" for the same reasons outlined above. Each attempts to answer different types of questions, using different methods and assumptions. Physical geology lends itself to making repeated observations and tests in real time, while historical geology is based on making inferences about the unseen past based on a variety of assumptions, such as uniformitarianism. Assumptions of course are things that are believed to be true even if you cannot use science to prove that they are. Uniformitarianism is not science. Rather, it is a philosophical position held by many scientists. It is a popular, unproven assumption.

Operational science and historical science are both valuable, but they are different. Historical science tends to be more speculative in nature as we have circumstantial evidence about unobservable past events. That does not mean that we cannot form inferences about the past, but we cannot directly test our conclusions as we would in operational science. Attempting to draw inferences from a past, unseen event such as the origin of the universe relies heavily on what you already assume to be true. The issue then is not science v. faith when it comes to disputes regarding historical science. Rather, it is a battle of conflicting assumptions and there is "faith" required regardless of what those assumptions are. In astronomy, there may be a consensus among mainstream scientists—based on their shared philosophical assumptions (their shared faith) – that the universe originated from the Big Bang, that the Earth's origin is best explained by the solar nebula model, and that life was the result of pre-biotic evolution. However, consensus does not mean something has been proven to be true.

1.3 Worldview and Bias

Like the word "science" there is no single, universal definition for the term "worldview," but most can agree on some broad descriptions. How you make sense of life on a day-to-day basis is determined by your worldview. You can think of your worldview as the metaphorical lenses through which you see things. Your worldview is the interconnected system of beliefs that steers your thoughts and deeds and interprets your experiences and observations. You might also think of your worldview as your own unique sets of biases. In short, everyone is biased, and the sources of that bias are the interconnected beliefs about reality, identity, values, and purpose (to name a few).

Despite the spectrum of possible worldviews, consider two extremes. If, like Nicholas Copernicus, Johannes Kepler, and Isaac Newton, your worldview is centered on a belief in the God of the Bible as the creator and sustainer of the universe, then how you view the world—including science—will be skewed accordingly. Science becomes a tool to better understand the created world, and doing science becomes an acceptable form of worship. On the other hand, if, like Neil DeGrasse Tyson or Richard Dawkins, your worldview forbids any acknowledgment or acceptance

Figure 1.9 A person's worldview will determine how they view the universe.

of the supernatural, then how you view the world will be quite different. This will be most obvious in the historical sciences.

There is no denying that among secular scientists the prevailing worldview aligns more with Tyson and Dawkins than with Kepler and Newton. Historical revisionists would likely argue that we are simply more "enlightened" today thanks to the continuous advances in science. Surely if Kepler and Newton knew what we know today, they too would have abandoned any sort of religious belief. However, the actual history of science does not support such a narrative. What is more relevant than any advances in science is the change in worldview. Honest scientists acknowledge as much. For example, consider this quotation from evolutionary biologist Ernst Mayr: "the Darwinian revolution was not merely the replacement of one scientific theory by another… but rather the replacement of a worldview, in which the supernatural was accepted as a normal and relevant explanatory principle, by a new worldview in which there was no room for supernatural forces." You do not have to be a creationist to recognize the importance of worldview, but most secular scientists never discuss its role in their work. Some will be quick to discuss the dangers of a worldview that is not their own, but rarely will you find any self-critique of their own bias.

When it comes to historical science—the origin of the cosmos, the origin of the Earth, the origin of life—the consensus that is reflected in most textbooks, journals, and classrooms comes from a worldview of naturalism. Put simply, naturalism says that all there is and all there has ever been at work in the universe is natural law. Importantly, if all that exists is "nature" then there is no such thing as the supernatural. Two other terms that are sometimes used to describe this position are materialism and reductionism. Materialism claims that all there is or has ever been is matter and energy. Reductionism goes a bit further, claiming that even very complex systems, including life itself, can be reduced to chemistry and physics. For example, reductionism would say that consciousness is nothing more than the product of brain chemistry. Do not miss the point here. Naturalism, materialism, and reductionism are all referring to sets of unproven assumptions. They make philosophical claims, not scientific claims.

In astronomy, as in any of the various branches of science, attacks on theism or even the simple suggestion of "design" in the universe are the result of bias, not objectivity. Everyone is biased. Everyone has a worldview. Science is no more biased than the hammer in your toolbox, but every scientist is. The history and philosophy of science do not support the current demand for unyielding naturalism. The Big Bang theory, the nebular hypothesis, and pre-biotic evolution may be the consensus views of origins, but the question to wrestle with is whether or not this is the result of good science or the dogma that springs from naturalism. Christians need not be intimidated or bullied into believing that faith in God as the Creator, and faith in the authority of scripture somehow makes them anti-science.

Late in the nineteenth century, progressive (revisionist) historians began to popularize the "conflict model" of science and faith. Science and religion had always been at war, the new story said, despite all the evidence to the contrary. Progress would depend on abandoning religion so that science could flourish. Late in the twentieth century this rewriting of history would be parroted by the likes of Carl Sagan and Stephen Hawking, and today most will mindlessly accept the notion as being true. The reality is that being a Christian does not make you less of a scientist, just as being a naturalist does not make you less religious in your beliefs. Rather than framing conflicts over origins as "science v. faith," a more honest assessment would be to recognize the role of competing worldviews as the source of debate.

1.4 A Second Voice

What you have read so far has been an attempt to briefly introduce you to important ideas that rarely make it into the pages of astronomy textbooks. It is an incomplete attempt at best but think of it as "putting our cards on the table," as the saying goes. These first pages reflect the thoughts of someone whose background has been in science education. What follows now is another perspective, one from a field scientist who also happens to have a Ph.D. in philosophy of science. The hope is that getting multiple, complimentary perspectives will help you recognize the big ideas and make the important connections before we move on to our topics in astronomy.

Charley Dewberry: Science and Philosophy

Dr. Charley Dewberry spent many years working in the streams and rivers of the Pacific Northwest, doing research on stream ecology and salmon restoration. Besides being a professional scientist, Dr. Dewberry is also a philosopher and an instructor at Gutenberg College in Eugene, Oregon. What follows are key excerpts from his book, *Saving Science*. While you may notice several references to salmon (his research area), the book addresses the general nature of science and several of the most common and perhaps dangerous misconceptions about science. Here you will read Dr. Dewberry's critique of the claims that (1) science is a privileged form of knowledge, and (2) science is objective.

Figure 1.10

© Krasowit/Shutterstock.com

The Objectivity of Scientific Information

Science is currently defined as a method of investigation that, if carefully followed, yields "objective knowledge," results that are more objective and more certain than could be obtained by other means of investigation. Science's exalted or privileged status results from its unique method of inquiry, a method of investigation often termed "hypothesis testing."

The scientific method, or hypothesis testing, is a mechanical process, and it is the mechanical workings of the method that supposedly ensure the outcome. This mechanism gives science the presumed objectivity, detachment, and certainty denied to other means of knowing because the outcome does not depend on any attribute within the scientist. Indeed, the method aims to detach the scientist from the test, negating any belief or any skill that a particular practitioner has.

Given this view, science could even be done by a machine. Samples could be placed into the machine, the outcome could be a dial reading that could be compared to the expected result, and statistics could be used to decide whether the results supported or refuted a particular hypothesis. This detachment and resulting objectivity are the basis for the claim that science alone deals with "facts."

Figure 1.11 Bronze of Sir Karl Popper.

The current view of science has largely come to us from the two philosophies of science dominant during the first half of the twentieth century, *logical positivism* and the philosophy of Karl Popper. From logical positivism come the ideas that scientific knowledge is the only form of knowledge and that science is objective measurement; that is, we can only have knowledge about elements that we can quantify, and our measurements must be neutral with regard to any other factor. From Karl Popper comes the view of science as hypothesis testing, and specifically, testing hypotheses in order to rule out possibilities by falsifying theories; that is, scientists, assuming that no one could ever prove a theory's truth absolutely, set up controlled experiments to try to show whether a theory is wrong. These two philosophies (which we will examine in more detail later) led to the idea that the unique value of the knowledge gained by the scientist is the objectivity and detachment by which the knowledge was gained.

Given this understanding, we can justifiably ask, who could be better at proposing and assessing salmon restoration strategies than the scientist, the one with access to a higher kind of knowledge than anyone else? Before answering this question, however, we must ask another: On what basis can a scientist make judgments and speak about what is true? Activities such as making judgments, diagnosing, evaluating, and proposing solutions all require skill, and yet the very method that ensures the unique objectivity and detachment of scientific knowledge—the mechanical nature of the process—requires that the scientist's skill be irrelevant to the results.

Just as a cook would follow a recipe in a cookbook, scientists mechanically follow the scientific method, which sets aside all experience as bias to be removed from the process. This is the basis of science's greater claim to objectivity. Reproducible results are a fundamental characteristic of science; scientists assume that *any* practitioner who adheres to the same method will get the same results, just as any cook who follows a recipe should get the same results. It is the mechanical working of the method—not any attribute of the scientist—that is important.

If adherence to the scientific method is what gives scientific knowledge its greater privilege, then the scientist has no superior basis for making any scientific judgments. Scientists become, in essence, unskilled workers. True, they are knowledgeable workers; they are disciplined workers. But the supposed virtue of the method, the superiority of scientific knowledge, arises from the fact that the results do not depend on the individual, biased attributes that the exercise of skill demands. Making judgments is a skill, but the supposed superiority of scientific knowledge depends on skill being irrelevant. Thus the dilemma: how can an unskilled worker use skills? To help unpack this dilemma we need to take a careful look at the nature of skills.

The Nature of Skills

Because the scientific method is designed to work against our everyday experience, one might argue that following it carefully requires a great deal of skill. While I agree that doing good science within the framework of the scientific method takes careful work, I would not call it skill. To make such a claim is to misunderstand the nature of skills. Following the scientific method, like following a recipe in a cookbook, is very different from performing skillfully. Following a recipe involves paying close attention to each step in linear order. Performing a skill involves much more.

A person learning any new skill— whether it be shooting baskets, riding a bicycle, playing a musical instrument, identifying an organism, or diagnosing a patient's health—faces a thousand "rules" that he or she must follow simultaneously. For example, when a boy learns to ride a bicycle, his parent tells him everything he needs to do. Initially the boy focuses on one or more of the rules, but his attempts to ride are awkward because he cannot follow all the rules at once. As he practices (i.e., as he focuses on doing particular rules), he gets more proficient. In time, following

Figure 1.12

© Denys Niezhientsev/Shutterstock.com

the rules becomes second nature to him, and he learns to follow a number of unarticulated rules as well. Eventually the boy no longer focuses on the rules; he just rides the bike. At this point, he has the minimum ability to perform the skill of riding a bicycle. In some sense, he is following all the rules, but he is not focusing on any one of them; he is just riding the bicycle. If the boy focuses on the rules again, then he has ceased doing the skill and is only practicing. Performing a skill, then, depends on the practitioner; it is an act of both "knowing" and "doing."

A skill is an achievement, which a person accomplishes by observing a set of rules he or she did not previously know. Now, the boy riding the bike does not know *all* the rules necessary to ride. He cannot articulate how he rides. He does not know how he balances. He does know that if he is falling to the right, then he must turn the handlebars to the right; but he does not know that turning the handlebars results in a centrifugal force that pushes him back to the left and keeps him from falling. Even though the boy does not know the explicit "rule" behind the maneuver, he can execute it without conscious thought because he has achieved the skill of riding.

Scientific Judgment and Skill

As we have seen, skill depends on a practitioner's personal knowledge of how to do something. In contrast, the goal of the scientific method is to eliminate personal knowledge—that is, to detach human biases from any results—in order to achieve the objectivity that supposedly makes empirical science unique and gives it greater standing. Thus, those holding the current view of science would labor slavishly following the prescribed method (much as a cook would follow a recipe) "science" and performing skillfully "art."

There can be no art in science—or so everyone seems to be saying. In order to gain objective knowledge, science must be a strictly mechanical process that precludes any skill. By thus elevating objective knowledge, we remove personal knowledge (including skills) from our accepted body of knowledge. Only the objective, peer-reviewed work remains.

Figure 1.13 Science is not a mechanical process.

If science is the mechanical process I have described, then no judgment exists within science. Only choices that can be determined mechanically by the scientific method are acceptable. Furthermore, any reference to scientific judgment is without content because making a judgment is a skill; and skill, belonging as it does to the realm of personal knowledge, is an art. When a person makes a judgment, he or she cannot construct a complete decision-matrix to determine the results mechanically because he or she is not conscious of all the factors that go into making the judgment nor how those factors are weighed. If we say the scientific method yields the only acceptable kind of knowledge, then we are saying knowledge has nothing to do with judgment. In fact, we are ultimately saying that knowledge has nothing to do with intellectual reasoning. The scientific method is based on empirical tests using statistics, and the purpose of using statistics is to remove any intellectual reasoning, including judgment, from the process.

We have come to the absurd conclusion, then, that making scientific judgments, which is an exercise of skill, are best made by those whose work excludes skill as a criterion. We can clearly see the absurdity of this position by making an analogy to medicine. When we are sick, do we go to the medical researcher or to the general practitioner for a diagnosis? The medical researcher is a scientist who tests hypotheses and does statistical analysis on the results and who, therefore, supposedly has superior, objective, unbiased knowledge gained through a mechanical process. And yet it is not the medical researcher but the general practitioner to whom we take our aches and pains.

Diagnosis is making a judgment, a skill. The basis upon which the general practitioner makes a judgment is the years of experience he or she has had making judgments—that is, his or her personal knowledge. Although the researcher may know all the literature on a particular topic—that is, the researcher has objective knowledge—he or she may not have made a single diagnosis since entering the research-side of medicine, and, therefore, the researcher likely does not possess the general practitioner's skill of general diagnosis. In fact, doing science, which detaches skills from the process, works against the development of skills. Over time, the general practitioner's diagnostic skill increases, but the researcher's diagnostic skill erodes, which is why we go to a general practitioner rather than a medical researcher when we are sick.

Analyzing a patient is very different from analyzing the statistical outcomes of various tests. The general practitioner and the medical researcher might rely on the same tests, but they evaluate them differently. We have not tried to replace the general practitioner with a written mechanical decision-matrix, nor do we ask for the one the practitioner used to determine our diagnosis. In fact, he or she could not construct a complete decision-matrix because the practice of diagnosing the health of a patient can never be reduced to a mechanical process.

Figure 1.14 Making a medical diagnosis is a skill.

To claim that only the medical researcher has any knowledge about medicine or that only the researcher is qualified to make judgments in medicine is the death of medicine. Such a claim denies the role of the general practitioner and denies that personal knowledge is true knowledge. The method of the general practitioner is not scientific; the practitioner's work is based on his or her skill, and skill is outside the realm of science. Yet, if knowledge in medicine is limited to the objective knowledge of the medical researcher, then medicine can no longer diagnose because diagnosis is a skill and not objective knowledge; and furthermore, judgment cannot exist within medicine because making judgments is a skill based on experience and therefore lies outside of science.

With our medical analogy in mind, let us look at salmon restoration. Who is qualified to diagnose and prescribe in this situation? According to the prevailing view, only the scientist is allowed to provide the facts and to make judgments about the recovery of salmon. The superior, objective, unbiased knowledge is the knowledge found in the peer-reviewed science literature. Therein lies the problem. Diagnosis is a skill. But to the extent that scientists have faithfully practiced the scientific method, they have spent their lives setting aside skills, including judgments. They are, in fact, handicapped when it comes to making judgments. On the other hand, people who have spent their entire lives working and living with salmon—those who have developed skills that could be the basis for judgment—are viewed as having only personal knowledge about salmon. According to the prevailing view, those who have had the least opportunity to develop judgment skills are seen as more qualified than those who have had the most opportunity to develop those skills.

Scientific Judgment and Experience

As we saw in the example of the medical doctor, making judgments is a skill, and skill is developed through experience. Experience, then, is a foundational requirement for making good diagnoses. As we have also seen, however, the scientific method tends to eliminate skill as a factor; thus having a body of peer-reviewed work may not indicate that a researcher has any experience making judgments in the field.

In fact, the problem is potentially more serious than I have indicated. Since the scientific method is designed to detach the scientist from the process, how better to do that than to have other people collect the data? Consider the following scenario; it is not unusual. For a major research project, the lead scientist writes a grant proposal in which he or she outlines a hypothesis or what is to be done, the methods that will be used to collect the information, and the statistical methods that will be used to analyze it. When the scientist gets the grant, he or she hires technicians to collect the data. The scientist then prepares data forms and spends time in the field—often a day or less—showing the technicians how to collect the data and enter the information on the data form. This may very well be the only time the scientist sees the project in the field. The "real" scientific work is setting up the test properly, which the scientist has done, and then analyzing the data statistically and writing the report. The scientist, however, may not even do the statistical analysis; he or she may only write the final report. And if the scientist has good technicians, the technicians may even draft the manuscript.

Since the method of science—not the scientist—is important, whether a scientist actually does any of the work on a project is irrelevant. When all the lead scientist does, then, is write the grant proposal and get the money, on what basis does he or she speak for the results or make judgments about it? Indeed, the scientist has very little basis on which to speak about the project because the basis on which someone can make judgments is experience, and the primary researcher often has little experience with the data collection.

Understanding this implication of the objectivity of the scientific method has helped me understand two of the more troubling experiences of my graduate education in fisheries and wildlife at Oregon State University. The first experience was seeing the emphasis on statistics over experience. On the one hand, I had always assumed that because a scientist's judgment is built on years of field experience, field experience was the most important element of good science. I therefore emphasized field experience as part of my education; I never missed an opportunity to spend time in the field, and upon hearing the results of a recent experiment, I would sometimes respond that it did not make sense given my experience. I would be told, however, that my experience was only anecdotal; it was not factual or scientific. On the other hand, my graduate committee wanted me to take more than half of my doctoral program in statistics. I fought it, but still I took more than two years of it. My point is not that statistics do not have a role to play in science, but that the role is minor compared with the experience and skill gained in the field. When I was in graduate school,

Figure 1.15 It takes field experience to become a true expert.

a student could complete the course work in stream ecology without taking a single field course. (Several courses included a field trip, but that hardly constitutes field experience.) To propose taking field courses and no statistics, however, would not have passed the laugh test. Completion of a thesis project was the only field experience deemed necessary for a scientist; experience and personal knowledge were not considered necessary for scientific knowledge.

The second experience that troubled me in graduate school was seeing how little actual field experience, or personal knowledge, many researchers in salmon-related fields had and then discovering that this lack of experience was not seen as a problem. Today some of these scientists with little field experience are on scientific panels making critical evaluations and decisions about salmon recovery. Some nationally and internationally known scientists who publish literature on salmon have not conducted extensive fieldwork since they completed their degrees or post-graduate work, and some cannot even identify the fish in the field. This is not the case for all scientists on the panels, but a number of them have little field experience with salmon. In fact, the ability to identify juvenile and adult salmon in the field is not a prerequisite for membership on these committees. My point is that no one sees this as a problem.

Judgment and Personal Knowledge

Thus far we have seen that assuming science ought to have a privileged position in salmon recovery leads to absurd conclusions: the diagnoses of those whose work de-emphasizes skill are to be preferred to those who have skill, and the diagnoses of those whose work de-emphasizes experience are to be preferred to those who have experience. Now, let us return to our examination of skills for a moment and the example of the boy riding a bicycle to discover the final absurdity of assuming an exalted view of science: the bizarre view of knowledge that must result.

Figure 1.16

Physicists and engineers have determined the specific physical laws (the objective knowledge) necessary for riding a bicycle. But what if a particular physicist who can articulate these laws cannot ride a bicycle? Does this physicist have a greater understanding of bicycle riding than the boy who, using personal knowledge, just rides the bike? While I would acknowledge that the scientist knows something that the boy does not know explicitly, I would deny that the scientist's "objective" knowledge of physical laws is a higher or more desirable kind of knowledge. And, therefore, I would also deny the logical positivists' claim (which pervades the prevailing view of science) that objective knowledge is the only form of knowledge.

To claim that objective knowledge of the laws of bicycle riding is the *only* knowledge of bicycle riding is wrong. It is absurd to claim that our non-riding physicist has knowledge of bicycling riding but that our boy bicyclist has no knowledge of bicycle riding just because he cannot articulate certain physical laws. The notion of objective, more certain knowledge is largely irrelevant in this case; the relevant knowledge is the skill of bicycle riding, or personal knowledge. The physicist has a very restricted and not very useful knowledge of bicycle riding; he or she has no skill. The bicyclist may not be consciously aware of the physical laws governing bicycle riding (although he must "know" them in some sense in order to ride), but he can use them; he has skill.

Likewise, to accept only the research scientist's objective knowledge when considering issues in salmon restoration is wrong. Doing so relegates the field experience (the personal knowledge) of workers and fishermen who have spent major portions of their lives working with salmon on a daily basis to opinion, to second-rate anecdotal information. Such an exalted view of empirical science and its objective knowledge cripples our restoration efforts because the research scientists often lack basic experience with salmon and the resulting skills that are crucial for making intelligent judgments. In fact, as we have seen, the scientific method itself mitigates against the development of that experience and skill.

According to the view of science accepted by almost everyone, experience and time in the field are irrelevant; these are biases best eliminated from the scientific process. An implication of this view would seem to be that the best scientists are those with the least experience—they have fewer biases to overcome. I do not agree.

How did the widespread view of the unique objectivity of scientific knowledge come about? I am aware of two versions of the argument for the certainty and objectivity of scientific knowledge—what we might call the "strong" and "weak" forms of the argument. The logical positivists made the strong form of the argument, and I call it "strong" because the positivists argued for the total objectivity of scientific knowledge; they viewed scientific knowledge as uniquely certain because it was built totally on objective facts and not at all on subjective values. The weak form of the argument does not go so far—it acknowledges that values do enter into the scientific process—and yet it still grants scientific knowledge a unique status because the *testing* of hypotheses is value-free. This form of the argument was made by Karl Popper. I will critique both perspectives in turn, first that of the logical positivists and then that of Karl Popper.

Critique of the Logical Positivists

According to the strong form of the argument for the unique objectivity of scientific knowledge, science has a unique status because scientific knowledge alone is objective—that is, value-free—and the results generated by slavishly adhering to the scientific method are objective facts. This view of scientific knowledge was the cornerstone of most versions of logical positivism, the most fashionable philosophy of science throughout most of the twentieth century. The positivists believed science alone dealt with facts while all other endeavors to acquire knowledge were fraught with subjective values. Therefore, many positivists believed that scientific knowledge was the only form of knowledge; all other forms of "knowledge" were mere opinion.

In the following section, I will present three criticisms of the view of science advocated by the logical positivists. The first two critiques are relatively straightforward and simple. The third critique is more involved, and I must develop it in some detail.

First Critique: Common Practice

According to the logical positivists, science is "value-free," but in actual practice it is not. For example, when I am asked to review manuscripts submitted for publication, one of the criteria for evaluating them is always the question, "Is a paper interesting or significant?" This criterion is a value, not a fact. Using a value judgment to screen submitted papers thus smuggles values into the scientific literature. Furthermore, scientists themselves pick their hypotheses based on values. Without values there would be no criteria for selecting a particular hypothesis; scientists would end up randomly investigating the most common phenomena on the planet, only rarely landing on an interesting, valuable hypothesis.

A truly value-free science implies that no hypothesis is inherently better than any other; no criteria exist to rank or prioritize hypotheses, and therefore all hypotheses are equally valuable. A truly value-free science, then, is analogous to a camera that sees everything and nothing. Only the photographer who values some things over others can communicate something in a photograph. So, for example, in a truly value-free science, investigating whether streams whose names begin with the letter "S" are more productive for salmon would be as worthwhile as any other hypothesis, and the only criterion someone could use to evaluate any manuscript would be whether or not the reported results were achieved by mechanically following the scientific

Figure 1.17

method. In the real world, however, even if the testing of "The-Letter-S" hypothesis adhered strictly to the scientific method, reviewers would rightly reject the manuscript reporting the results on the basis of values, namely, common sense and experience. Common sense and experience are not facts in the sense required by a positivist view of science. Indeed, the purpose of statistical procedures in science is, in part, to remove these very human elements; and yet both common sense and experience are smuggled in from outside the articulated scientific method. Thankfully, when it comes to selecting hypotheses, no scientists really attempt to follow the demands of a truly value-free science (even though they may pay lip service to it) because it would only lead to a morass of isolated facts and tests of mostly irrelevant hypotheses.

The science *practiced* by individuals (e.g., those working on salmon recovery) does not actually separate facts from values; science as practiced does not rely only on facts. Furthermore, no scientist could follow the articulated method in a way that would lead to objective knowledge, and the scientific literature itself is clearly not value free.

Second Critique: Making Connections

The logical positivists never specified the rules of induction—that is, how facts are linked together. This "connection" problem has existed in modern empirical science since its conception by Francis Bacon and others. It is one thing to come up with the facts; it is another to link them into some kind of coherent picture or theory. The positivists saw everything as either a fact or a value. Given this view, therefore, the linkages must be values because they are not facts; they are not just observed. Furthermore, since there are no established rules for the process of linking facts into a coherent picture, it cannot be objective and detached and totally reliant on facts.

One might think to get around this problem by arguing that there is no picture or theory that links and organizes the facts; there are only isolated facts, and science is simply what accumulates them.

Figure 1.18 Science is not simply the accumulation of facts.

Many who hold a positivist view of science favor this position that reduces a theory's status to nothing more than a summary of the facts. This argument, however, does not describe the real world. In the real world, human beings—whether we want to or not and whether we admit it or not—link facts together to form a coherent, whole picture. To do so, we always rely on our experience and prior beliefs, and this effort comes completely from outside the scientific method. In practice, then, objective scientific facts are always linked together in theories, and since those theories are not "facts," they must be values.

Third Critique: Hume's Epistemology

According to the logical positivists, scientific knowledge is uniquely certain because it is built totally on objective facts. But there are no such things as objective facts. Because this criticism of the positivists' view of science is more technical, I must first provide some background. To the positivists and other tough-minded empirical scientists, observation is just opening one's eyes and looking. Facts are simply the things that happen—hard, sheer, plain, and objective; someone sets up an experiment and gets a number. The positivists derived their epistemology explicitly from philosopher David Hume; they agreed with Hume that matters transcending human experience have no meaning and even if they had, they could not be shown to be true. Hume's empiricism can be summed up in two propositions.

Figure 1.19 Philosopher David Hume.

© Prakich Treetasayuth/Shutterstock.com

1. All our ideas are derived from impressions of sense or feeling. That is, we cannot conceive of things different in kind from everything in our experience.
2. A matter of fact can never be proved by reasoning *a priori*. It must be discovered in, or inferred from, experience.

According to this view, facts are the simple impressions of sense that arise in the mind from unknown sources. Once we have perceived the simple facts, then we can arrange and rearrange them, but initially we experience them passively; the mind is not active during this stage; the impressions just appear. For example, when I experience the color red and a particular shape, I call the combination of these two simple impressions a "strawberry." To claim something is a "fact," then, is to claim it can be traced back to impressions. A fact's empirical pedigree therefore determines whether or not we are justified in believing it; if there is no impression, then there is no fact.

A number of important implications arise from accepting Hume's epistemology. We only know the ideas in our mind because they are all we can experience. Therefore cause-and-effect refers to ideas in our mind, not something external to us. For example, we think the Sun will rise tomorrow because that thought is a habit of our mind; there are no physical laws or other explanations for the sunrise, and the Sun is just as likely not to rise tomorrow. What we mean by causation is that two ideas appear one after the other in our minds (what Hume calls "constant conjunction" or "correlation"), and that is all we know; we can never speak of causes as if they were in a world outside of our minds. We do not even know if a world exists outside of our minds or what it might be like. Likewise, other people and our own bodies are just ideas in our minds; we have no other knowledge of them or us. We do not know if rivers exist, or salmon, or anything else outside of our minds.

The positivists were aware of these implications of Hume's epistemology, and they accepted them. I see four problems, however, with the positivists' view of facts as objective and empirically justified. None of the following observations are uniquely mine; I suspect they have been made many times in many different circumstances, and I will acknowledge my debt if and when I remember who first called them to my attention.

Problem One
An Erroneous Theory of Observation/Perception

The first problem with the idea of facts as objective and empirically justified is the faulty view on which it rests: Hume's (and others') erroneous theory of observation/perception. If Hume's theory of observation/perception falls, so does the claim for "objective" knowledge. The basis for the radical empiricist's claim is that impressions are facts. Impressions just happen (they either occur or do not), and because nothing mediates them, they are always reliable. If, therefore, two researchers run the same experiment, they should get the same dial readings; the dial readings either are the same or they are not, and nothing in the researchers can affect the outcome. Since the 1950s, however, philosophers of science have found Hume's observational/perceptual theory to be in error, and two examples they cite that are most damaging to Hume's theory are the cases of ambiguous figures and inverted lenses.

Figure 1.20 A simple example of an ambiguous figure—what do you see?

In his book, *Patterns of Discovery,* N. R. Hanson presented a number of ambiguous figures, which some people see as one thing (e.g., an antelope) and others see as something else (e.g., a rabbit). Some people can only see one or the other figure. Some can move from one figure to the other. In all cases what strikes the retina is the same, yet the objects the observers see are clearly different. The observers are not aware of seeing a series of lines, which they would all see the same; they are only aware of seeing a figure, and those figures differ. Hume's observational/perceptual theory, however, is based on the idea of bare "observation"; that is, the series of lines is perceived, and then the mind interprets it. For Hume, perception/observation is a mechanical process: first the impression occurs, and then the whole or parts of the impression can be copied, moved, and stored within the mind. The observers, however, do not report seeing the lines; they only report seeing figures. One could argue that the lines are seen subconsciously and the observer moves quickly to interpretation; however, this does not solve the problem. From an empirical point of view, what is the difference between two observers not being aware of a base perception and there being no base perception at all? Empirically, these two cases cannot be separated. These examples of ambiguous figures call into question the assumption that observation is a simple, mechanical process not mediated by past experience or other beliefs. Said another way, they call into question the idea that facts are just observed, that they just "happen" to the observer.

The case of the inverted lenses is even more damaging to Hume's observational/perceptual theory. A pair of glasses with lenses that invert an image on the retinas is placed in front of an observer. At first the observer is completely disoriented; with a little time, however, the observer learns to see normally. In fact, with a little time, the observer is not consciously aware that the image is inverted on his or her retinas; the observer sees the image just as someone who is not wearing the glasses would see it. At this point, the sense data are different for the observer wearing glasses and the one not wearing glasses, but the two observers see the same thing. If we passively perceive the sensations first, as Hume's theory states, why does the observer wearing the glasses not see the inverted image and then have to interpret it actively each time?

Philosophers of science have rightly rejected the Humean theory of observation/perception for about fifty years. The radical empiricists' claim that facts are just passively observed is false. Rather, the mind is active during observation/perception, and, therefore, what we know (all our accepted beliefs) affects what we see. The radical empiricist who wants to claim that facts are just observed objectively is wrong. There are no objective facts in this sense. While the radical empiricists see science built one objective fact at a time, their critics see each fact as "theory-laden," that is, already imbedded in a theory and experience. Furthermore, implicit in this criticism of Hume's theory is a rejection of the radical empiricist's claim that the nature of observation is mechanical. Rather, observation/perception is a skill that the observer has honed throughout his or her life. It is not a mechanical process; it is high art that can never be made objective. What individuals see depends on their experience.

Problem Two
An Erroneous View of Theory Change

The second problem with the idea of "objective" facts is closely related to the first, and philosopher Thomas Kuhn proposed it in the 1950s. While the philosophers discussed above saw facts imbedded in theories, Kuhn saw them embedded in "paradigms." Facts, according to Kuhn, are only understandable within the context of a paradigm.

From a positivist view, science progresses as isolated, value-neutral facts are accumulated. Once established, facts are the "unit of currency" for science; they do not change. And because progress in science is seen primarily as the accumulation of facts, a theory changes only when a new theory explains all the existing facts better than its predecessor. Science, therefore,

Figure 1.21

progresses linearly, which is why science can uniquely separate fact from values: it is based only on the facts, pure and simple.

Kuhn disagrees with the positivists. For him, theory change is not based primarily on the accumulated facts, and it is not the logical, rational, and mechanical process that positivists believe it to be; rather, it is a paradigm shift. Science is not a lone scientist working in isolation, but a community of workers with a set of shared commitments; and as a science matures, it develops a "paradigm"—that is, a set of questions, methods, and case examples that guide the community of workers in their work. But a number of cases are hard to bring into the existing paradigm, and thus ordinary day-to-day science is problem solving; the scientist's work is to figure out how to bring these difficult cases into the paradigm. The more difficult and protracted the problem, the greater the prestige the scientist who finally solves it gains from his or her peers. Over time, however, the number of difficult cases that do not fit into the paradigm increases, and a "crisis" arises. At this point, a new (usually young) worker from outside the current paradigm proposes a new paradigm, which has a new set of questions, new methods, and new case examples. Individuals embracing competing paradigms cannot even communicate with each other because they are "speaking a different language" (a different language game); they have "a different way of life," to borrow a phrase from Wittgenstein. According to Kuhn, therefore, theory change in science is like change in other arenas of life: it can be revolutionary.

In Kuhn's model, a new paradigm does not have to incorporate all the facts of the previous paradigm. Indeed, facts that were seen as critical and fundamental in the old paradigm may not even have standing in the new one. The accumulation of facts, therefore, is no longer the benchmark of science.

According to Kuhn, there are no criteria to decide logically and mechanically if the new paradigm should be embraced; there is no objective standard by which to judge competing paradigms. Scientists must base their decision to accept the new paradigm on other grounds. Somehow, though, a critical mass of workers accepts the new paradigm, the new paradigm replaces the old paradigm, and a new community of workers becomes the body of scientists.

In Kuhn's view, science has no unique status. It is like other endeavors of life, with no more ability to separate fact from opinion than any other endeavor. According to Kuhn, the claim made by the positivists and others to have objective knowledge is false. In this much I agree with Kuhn; later, however, I will discuss how I disagree with him.

Problem Three
An Erroneous View of Knowledge

The third problem with the idea of "objective facts" is the Humean view of knowledge it assumes—that is, human beings start with a "blank slate," and all their knowledge originates as impressions. If we start with a blank slate, however, how and on what basis could we respond to the first impression? We cannot know anything about it; there is no reason to think the impression is significant, no way to understand its meaning, and no reason to accept it. Given this blank-slate theory of knowledge, the radical empirical project could never get started.

Figure 1.22 Latin for "scraped tablet," but we use the phrase "blank slate."

To help make this criticism clear, think of the mind as analogous to a computer. If someone starts up a computer that has a blank slate—no Read-Only-Memory (ROM)—it will never work. The computer will not recognize the inputs from the keyboard nor have any idea how to understand the keystrokes. ROM tells the computer that inputs are significant and how to understand them, and without those initial instructions the computer is useless, Similarly, the human mind must have "instructions" for understanding its first impression; it is not the blank slate Hume posits.

Problem Four
A Philosophy Unworkable in Practice

The last problem with the concept of "objective" facts also relates to Hume's radical empirical philosophy that says we cannot know anything outside of our minds. Although I cannot prove it, I do not believe that any scientists involved with salmon recovery actually believe and accept the implications of Hume's theory. On the contrary, they believe that real streams and real salmon exist in the world and not just in their minds. They also believe that cause and effect is an attribute of the external world and not an attribute of the contents of their minds. And they believe that the external world and other people exist, as do their own bodies. Not for a minute do scientists live or practice their science holding the skeptical beliefs demanded by Hume's system.

The seemingly universal practice of scientists, therefore, denies the philosophical foundations for the positivists' view of science. The idea that scientific knowledge alone is certain because it deals with objective facts (other kinds of knowledge being less reliable because they deal with values) comes to us from Hume by way of the logical positivists. Yet the practice of scientists themselves implicitly rejects Hume's epistemology, the philosophical basis for the idea that science deals uniquely with value-free and objective facts.

Conclusion to the Critique of the Logical Positivists

We have seen that the positivists' view of science—that is, the strong form of the argument for the objectivity of scientific knowledge—is in error. Practicing scientists in fact often interject subjective values into their science; subjective values must enter into the process of linking facts into coherent theories; and Hume's epistemology, embraced by the logical positivists in developing their view of science, is theoretically and practically unworkable.

Critique of Karl Popper's view

Perhaps most practicing scientists would not recognize their own position in the strong form of the argument for the objectivity of scientific knowledge. They would agree that values enter the scientific process when a hypothesis is selected and the manuscript is peer reviewed. Yet they would still argue for a unique objectivity to the scientific method. And to do so, they would use the weaker form of the argument for the objectivity of scientific knowledge: namely, that the aim of the scientific method is to eliminate values from the *test* of the theory or hypothesis.

Figure 1.23

According to the major proponent of this view, Karl Popper, science is testing hypotheses, and the scientist's goal is to attempt to falsify theories by proposing bold hypotheses. In Popper's view, a theory can never be proven true; it can only be falsified. If a scientist finds even one case in which a theory is false, then the theory is not true. Currently accepted theories, then, have been tested but not yet rejected.

Popper's system contains a far more sophisticated view of the interaction of theory and fact than we found in the positivists' system. For the positivists, a theory is just a summary of the facts; it can be nothing more. For Popper, two theories can present two different views and both account for all the facts. The scientist's goal is to design a crucial experiment that can test between the two theories, and the scientist's motivation is to try to reject the currently held theory.

In the following section, as I did for the positivists, I will present three criticisms of the view of science advocated by Karl Popper.

First Critique: *Who Is a Scientist?*

Popper's view defines science in a way that would exclude many individuals I consider to be great scientists: Copernicus, Galileo, Kepler, and Newton, for example. None of these individuals primarily tested hypotheses; they were primarily restructuring our theories of the heavens. To define science in a way that excludes these individuals cannot be right.

Second Critique: *The Problem with Falsification*

Popper's method based on *falsification* of hypotheses or theories fails for the same reason that he claims that *verification* of hypotheses can never be certain. In his view, a theory can never be *verified*; any number of positive responses can never prove with certainty that a theory is true, because the possibility always exists that we have not examined a crucial case. But, in his view, *one* negative case proves that a theory is not true and that we should reject it; a theory can only be *falsified*.

Consider, however, the following case from the history of science. Astronomers were checking Newton's theory against the calculations of the outermost known planet at the time, and the orbit of the planet was not where Newton's theory predicted it should be. According to Popper's system, finding one case falsified Newton's theory, and the theory should have been rejected. Rather than rejecting Newton's theory, however,

the astronomers hypothesized that an undiscovered planet might be affecting the orbit of the outermost planet. They calculated where such a planet would have to be, and they did indeed discover a new planet beyond the known planets.

This example raises a crucial question: when do we know that a result that differs from the predicted result is actually a case of the theory failing and not the result of an unknown factor? There is no way to answer this question—just as there is

Figure 1.24

no way, according to Popper, to verify a theory. And so Popper's falsification criterion suffers from the same critique that he made against the verification system: in both cases the possibility exists that an unknown factor could change our conclusions. In the end, therefore, one false case does not necessarily falsify a theory.

Curiously, Popper anticipated this problem, but his answer was less than satisfying. He argued that, in theory, his empirical method exposes to empirical testing *every conceivable way* a system could be tested and, therefore, provides a means of preventing an apparent false case like the example above, but how can we be so sure that we have included and tested all *unanticipated* factors? Popper's answer does not really address this issue.

Third Critique: *Common Practice*

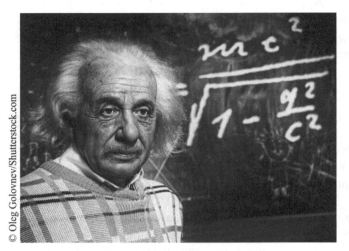

Figure 1.25 Albert Einstein

Popper's view seems antithetical to how we acquire and hold knowledge. We do not spend our time devising critical experiments to test all our cherished theories; we accept a theory unless some issue is raised that demands we revisit it. But even if we must revisit a cherished theory, the idea that we would naturally reject it based on one false empirical result—even from a crucial experiment—seems wrong. Popper values his empirical test too highly. It might be reasonable to shelve a result for a period of time, and negative empirical results do ultimately refute the truthfulness of a theory, but rejecting a theory is not always as easy or as mechanical as Popper claims.

To illustrate how Popper's view of testing hypotheses is too simplistic and mechanical, consider the case of D. C. Miller, which was brought to my attention in Michael Polanyi's book, *Personal Knowledge* (pp. 12–13). As is commonly known, Einstein based his theory of relativity in part on the results of the Michelson-Morley experiment. In a long series of experiments from 1902 to 1926, D. C. Miller and his associates repeated the Michelson-Morley speed-of- light experiments thousands of times with increasingly more sophisticated equipment, and yet each time they did the experiment their results contradicted the results predicted by Einstein's theory of relativity. Miller reported this in his presidential address to the American Physical Society in 1925. And yet, in spite of the fact that these experiments seem to falsify Einstein's theory, little attention was given to Miller's research. Physicists assumed that the results would be found to be in error at a later date. According to Karl Popper, this reaction was wrong; physicists should have rejected Einstein's theory immediately. But Popper has failed to understand the complex relationship between theories and fact, especially as that relationship is worked out in practice. To falsify a theory is not as easy or as objective as Popper thinks.

Conclusion to the Critique of Karl Popper

Just as we saw that the positivists' strong form of the argument for the objectivity of scientific knowledge is in error, we have now seen that the weaker form of the argument as made by Karl Popper is also in error. Although Popper (unlike the positivists) would acknowledge that values enter into the scientific process, he endeavors to preserve the unique status of scientific knowledge by insisting that the *testing* of hypotheses is value-free. Yet not only would Popper's definition of science exclude many great scientists, neither logic nor practice supports the idea that his method is uniquely objective.

Science is Not Uniquely Objective

In the minds of many, empirical peer-reviewed science has priority over "subjective" areas, such as field experience, historical research, creative theory formation, and so on. According to this view, the privileged role of science arises from science's unique ability to separate fact from value by means of the scientific method. As we have seen, however, neither the strong claim of the logical positivists nor the relatively weaker view of Popper holds up under scrutiny. No one can claim that science has a unique status because it deals with objective facts, because there are no such things as objective facts—that is, facts detached from any values. If the mind is active during observation, and observation is dependent on all our previously accepted beliefs, then there is no longer any foundation in which uniquely objective knowledge can take root. The scientist has no grounds to claim a unique certainty denied to other disciplines.

I do not believe, nor do I mean my critique to suggest, that empirical science is unimportant or indefensible. Empirical science plays an important role in science and, more specifically, in salmon recovery. Rather, my point is that the exalted status that radical empirical science has enjoyed is not defensible. My argument, then, is not against empirical science per se, but against a radical form that claims to be the ideal science and worthy of exalted status; neither one of these claims is true.

Appendix A JOHN WEST: Science & Culture

******

Sky Basics: Signs, Seasons, Days, and Years

Chapter 2

> *Then God said, "Let there be lights in the expanse of the heavens to separate the day from the night, and they shall serve as signs and for seasons, and for days and years..."*

—**Genesis 1:14**

2.1 Motions of the Stars

The Celestial Sphere

*As seen from Earth, the sky can be thought of as a gigantic sphere, centered on Earth, to which all objects seen in the sky are attached. This is called the *celestial sphere* (see Figure 2.1). Points on the celestial sphere are directly above the points on Earth, to which they correspond. The point on the celestial sphere directly above Earth's North Pole is called the *north celestial pole*, often referred to as the "NCP." The point above Earth's South Pole is called the *south celestial pole*. The line directly above Earth's equator is called the *celestial equator*.

To all Earth-based observers, the sky appears to be in the shape of a hemisphere or dome bounded by the *horizon*, as shown in Figure 2.2. Standing on a spherical Earth, which is comparatively large, an observer will detect very little curvature on the local scale, so we represent the local horizon by a flat, circular disk. The portion of the celestial sphere visible to an observer is said to be above their horizon, while the portion not visible is said to be below their horizon. The point in the sky directly above the observer is called the *zenith*. The compass points on the horizon—north, south, east, and west—are

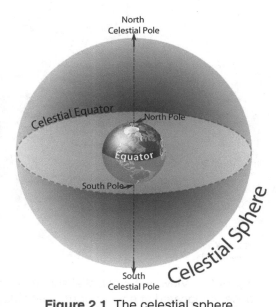

Figure 2.1 The celestial sphere.

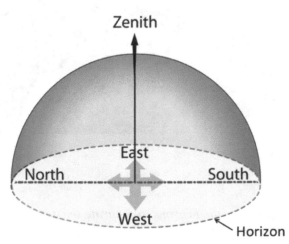

Figure 2.2 The sky's "dome" from an observer's point of view.

collectively known as the *cardinal points*. The *azimuth* refers to the angular distance along the horizon. An azimuth of zero degrees is north, 90 degrees is east, 180 degrees is south, and 270 degrees is west. An azimuth of 360 degrees is also considered north, completing the "circle."

Note in Figure 2.2, that the angle from the zenith to the horizon is 90 degrees. An observer's zenith will always be directly over their location on Earth, and they will, therefore, be able to see any part of the celestial sphere above their horizon in their sky that is within 90 degrees of their zenith. Angular measure from the horizon (zero degrees) to the zenith (90 degrees) is known as *altitude*.

Which half of the celestial sphere an observer will see depends on their local time and where on Earth they are located. For example, referring back to Figure 2.1, an observer at Earth's North Pole will see the north celestial pole at their local zenith. On their local horizon, 90 degrees below their zenith, they will see the celestial equator. Also note that an observer at Earth's North Pole will see the entire northern half of the celestial sphere. All of the stars and other objects that are north of the celestial equator will be visible to them. However, the southern half of the celestial sphere is *not* visible from the North Pole. An observer there will *never* see objects south of the celestial equator. All the objects in the southern half of the sky are always below their horizon. The opposite will be true for an observer at Earth's South Pole, who will never see the objects in the northern half of the celestial sphere.

Much of the world's population lives around halfway between Earth's equator and the North Pole. This region is called the mid-northern latitudes. *Latitude* is the measurement of how far north or south a location on Earth is from the Earth's equator.

As shown in Figure 2.3, an observer at a mid-northern latitude will look toward their zenith and see a point that is about halfway between the celestial equator and the north celestial pole. So, they will see the north celestial pole about halfway

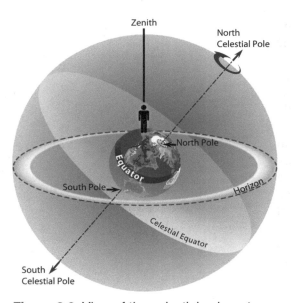

Figure 2.3 View of the celestial sphere to an observer located at a mid-northern latitude, about halfway between the Earth's equator and the North Pole.

up from the horizon in the northern part of their local sky as shown in Figure 2.4. They will see the celestial equator about halfway up from the horizon in the southern part of their sky.*

For anyone in the northern hemisphere, the North Star (Polaris) not only serves as a marker for direction (azimuth), but also for latitude. If someone sees Polaris at an altitude of 42 degrees above their horizon, it means they are at 42 degrees north latitude on the Earth.

Using good approximations of an object's altitude and azimuth in the sky to describe the location is known as the *altazimuth* or *horizon system*. Besides altitude, azimuth, and zenith, another useful reference point is the *meridian*, an

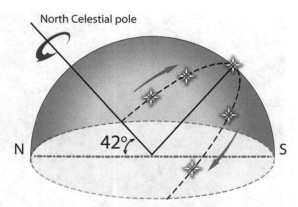

Figure 2.4 View of the sky for an observer at a mid-northern latitude.

imaginary line that runs from an observer's southern horizon (azimuth = 180 degrees), through their zenith, and down to their northern horizon. The meridian provides a useful reference point. For example, an observer can describe an object's position in the sky based on when that objects *transits* (crosses) the meridian. The altazimuth system is pretty simple for new observers to use, but keep in mind that the altitude of an object in the sky will vary depending on your position on the Earth.

Describing distances in the sky is done using degrees. It's not an actual distance that matters, but rather the *apparent* distance between objects that is of interest. For example, if an observer wanted to describe how far across the sky the Big Dipper stretches, the answer would be given in degrees. A simple system to try is to extend your hand out at arm's length. At arm's length, your index finger represents about 1 degree. Hold up three fingers boy scout style, and you have about 5 degrees, and the width of a closed fist at arm's length is about 10 degrees. Spread your hand as far as you can, and the distance from the end of your thumb to the end of your little finger is approaching 25 degrees.

Professional astronomers also measure angular distances and angular diameters of objects like the Moon, but with much greater accuracy and precision. Their measurements will likely include smaller fractions of a degree such as arc minutes and arc seconds, not to be confused with the terms minutes and seconds that are used to measure time. The "arc" is important.

*The *stars* seen in the sky are points of light that appear to be fixed on the celestial sphere. There are stars all around the celestial sphere, some brighter than others and some even appearing to form patterns. Long ago, many of these apparent patterns took on cultural significance, often representing characters in local mythological stories. Early in the last century, astronomers officially divided the celestial sphere into eighty-eight *constellations*. So, to a modern astronomer, the constellations are nothing more than well-defined, bounded regions of the celestial sphere, many of which have odd shapes so as to preserve, when possible, the traditional apparent patterns.* You can think of the eighty-eight constellations as a complete "map" of the sky. While it is still common for us to think of a constellation as a connect-the-dots creature or object, it is more accurate to think of a constellation as a region, not unlike the way we view Virginia on a U.S. map. In other words, when we talk about the constellation Orion (for example), we not only mean the stars that make up

Figure 2.5 The Summer Triangle is an asterism made from the brightest star of each of these three constellations.

Figure 2.6 A common trick for finding the North Star, *Polaris*, is to use the two outer stars of the Big Dipper's bowl as *pointer* stars. An imaginary line drawn out of the Big Dipper's bowl along the line of the pointer stars will point to Polaris, the star at the end of the handle of the Little Dipper (upper right).

the mythological figure, but the region around those stars within the boundaries set by the International Astronomical Union (IAU).

Many observers are familiar with other arrangements of stars known as *asterisms*. Asterisms are more unofficial patterns of stars as opposed to the constellations mapped out by the IAU. An asterism may be part of a constellation. For example, the Big Dipper, Little Dipper, and Orion's Belt are asterisms found in the constellations Ursa Major, Ursa Minor, and Orion, respectively. An asterism may also be made up of stars from multiple constellations. The Summer Triangle is formed by three stars that each come from a different constellation—Vega (from the constellation Lyra), Altair (from Aquila), and Deneb (from Cygnus). The Winter Triangle is formed by the bright stars Procyon, Sirius, and Betelgeuse, which once again come from three different constellations.

*The star that appears almost exactly at the north celestial pole, which would make it almost exactly above Earth's North Pole, is called *Polaris* or the *North Star*. There are common misconceptions that Polaris is the brightest star in the sky or that it is the closest star to our Sun, but it is neither of these. What is special about Polaris is its location. Being above Earth's North Pole, Polaris does not move in the sky and will always indicate which way is north. As the celestial sphere rotates, the rest of the stars shown in Figure 2.6 will move in circles around Polaris.

Their entire circle of apparent motion will be above the horizon for observers at mid-northern latitudes, so they are visible every night. Stars that are always above our horizon are called *circumpolar* stars.

Stars further from the north celestial pole will *rise* in the east and move across the southern part of the sky, where they reach their highest angle above the horizon, or their highest *altitude*. They then descend toward the west and disappear below the western horizon, or *set* in the west. Refer to Figure 2.4, in which the arrows point in the direction of the daily motion of the stars.* In other words, what an observer will see changes with their position on the Earth. Observers at the North Pole (with Polaris at their zenith) would see all of the stars as being circumpolar, while observers at the equator (with Polaris on the northern horizon) would see no stars as being circumpolar. U.S. observers then see both—some stars that are circumpolar and other stars that seem to rise and set. The exact observations someone can make on any given night depends on where they are on the Earth, and what time of year it is. If they live in Virginia (for example) there are some stars, asterisms, and constellations that they will see during the year that someone in New Zealand will never see, and vice versa.

Celestial Coordinates

*As previously stated, latitude is a coordinate indicating how far north or south of the Earth's equator an observer is located. *Longitude,* on the other hand, is a coordinate measure of east and west locations. Both of these coordinates are shown in Figure 2.7. So, the two coordinates that represent the location of an observer on the surface of the Earth are latitude and longitude.

The equivalent measurements of latitude and longitude on the celestial sphere, also shown in Figure 2.7, are called declination and right ascension. *Declination* is the measurement of how far north or south an object in the sky is from the celestial equator, while *right ascension*, like longitude on Earth, is a measurement indicating east and west locations on the celestial sphere. So, the two coordinates that represent the location of an object on the celestial sphere are right ascension and declination.*

Unlike the horizon/altazimuth system, this *equatorial* system uses coordinates that do not change with the position of the observer. For example, Sirius, the brightest star as seen from Earth, will have a declination of −16.5 degrees and a right ascension of six hours and forty-five minutes, regardless of where you live. Many astronomers like this system for that very reason, but it can be a bit of a challenge to the beginner.

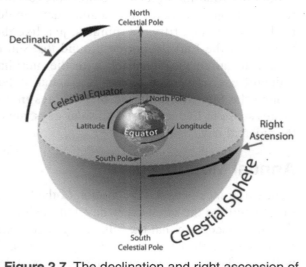

Figure 2.7 The declination and right ascension of an object in the sky are equivalent to the latitude and longitude of a location on Earth.

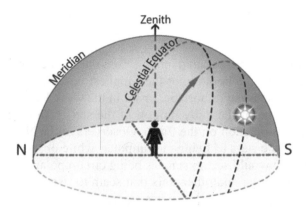

Figure 2.8 Daily motion of the Sun across the sky for mid-northern latitudes.

2.2 Motions of the Sun
Daily Motion

*As shown in Figure 2.8, the daily motion of the Sun is similar to the motion of the stars. The Sun rises more or less in the east, and its altitude increases until it is directly above due south, and then its altitude decreases until it sets more or less in the west. The line from due south through the zenith to due north, and dividing the sky into an eastern or rising half and a western or setting half, is called the *meridian*. The time that the Sun crosses the meridian is defined as local *noon*. The amount of time between two "noons" is our definition of the time measurement *day*. The hours before the Sun reaches the meridian are called *ante meridiem,* a.m. being the familiar abbreviation for the morning hours. The *post meridiem* hours, abbreviated as p.m., are the hours after the Sun has crossed the meridian.*

The 24-hour period between successive transits of the meridian by the Sun is known as one *solar day*. However, if we tried to use the stars to tell time based on successive transits there would be a noticeable issue. The time between successive transits of the meridian for a *star* is twenty-three hours and fifty-six minutes, what astronomers call a *sidereal day* (in Latin, "sidereus" means starry). The reason these two "days" are different is because each day, the Earth is not only rotating on its axis but also revolving around the Sun. If the Earth could magically rotate without also revolving, a solar day would not be twenty-four hours, it would be twenty-three hours and fifty-six minutes, just like the distant stars suggest. Think of it this way—because of the Earth's motion around the Sun (revolving), exactly one rotation of the Earth would not align the Sun back with the meridian. The Sun's position appears to shift about 1-degree each day, even though it is actually the Earth that is moving (360 degrees in one trip around the Sun over a period of 365 days in a year; that's about 1-degree per day). While it makes sense to live life based on solar days, anyone studying the stars needs to also be aware of the sidereal day.

Annual Motion

*The difference between the motion of the Sun and that of the stars is that a star's path through the sky will be the same every day, but the Sun's path is slightly different each day. It is a common misconception that the Sun always rises exactly due east and always sets exactly due west. If you observe the rising and setting positions and noon altitude of the Sun on any given day, you will indeed not notice much change in the next day or two. However, after about a week or more, you will notice differences. The rising and setting positions along the horizon and noon altitude will change.

As can be seen in Figure 2.9, when the Sun's rising and setting positions are at their furthest north of due east and west, respectively, the length of the Sun's path through the sky is longest. In the northern hemisphere, this occurs on or around June 21 every year and is called the *summer solstice.*

Throughout the days of summer, the Sun's path gradually moves southward and the noon altitude gets lower. Eventually, by about September 21, the rising and setting positions reach *exactly* due east and due west, respectively, and the Sun is, therefore, above the horizon and below the horizon for twelve hours each. This is known as the *autumnal equinox.* The word "equinox" means equal night, meaning the daylight hours and nighttime hours are equal. Note that since the Sun's altitude is getting lower, the Sun is heating the ground less efficiently and the weather in the fall becomes cooler.

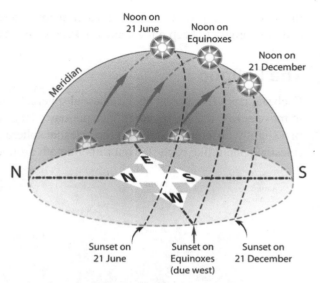

Figure 2.9 Differences in the Sun's daily path through the sky at different times of year.

As the year progresses, the Sun's daily path will continue to move south, and the noon altitude will continue to decrease until about December 21. The day of the Sun's shortest path and its lowest noon altitude is called the *winter solstice*, the first day of winter. Winter is the time of year with the least daylight and coldest temperatures. After this date, the Sun's rising and setting positions will start to move north again, and noon altitudes will increase. There will be more daylight, and the weather will get warmer. When the rising and setting positions return to due east and west, the noon altitude will be the same as it was on the autumnal equinox. This is the *vernal equinox*, or first day of spring, which occurs on or about March 21.

As shown in Figure 2.10, when the Sun is high in the sky, its rays strike the ground more directly and heat the ground more efficiently, causing warmer weather. When the Sun is lower in the sky, the rays are much less direct. The same amount of solar energy is spread out over a larger area, so the ground is being heated less efficiently, resulting in colder weather.

The amount of time required for the Sun to completely cycle through its changing daily paths is our definition of the time measurement *year.* So, the year is a natural cycle of the Sun taking approximately 365.25 days to complete. The extra 0.25 day is

Figure 2.10 The Sun's rays striking the ground from a lower (left) and higher altitude (right).

the reason for leap years. Our calendar contains 365 days, so every four years, a 1/4 day is added at the end of February to keep our calendar in sync with the Sun.

The Ecliptic

Each star has a fixed point on the celestial sphere, like cities have fixed points on Earth. All stars appear to move across the sky together once per day as the celestial sphere appears to rotate. Since the Sun is in a slightly different position on the celestial sphere each day, the Sun requires not just a point on the celestial sphere like each star but rather a path around the celestial sphere. The path extending all the way around the celestial sphere and tilted at an angle of 23.5 degrees relative to the celestial equator, intersecting it at two points, is called the *ecliptic*. See Figure 2.11.

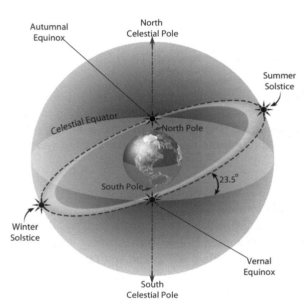

Figure 2.11 The ecliptic on the celestial sphere.

The points on the ecliptic are named for the northern hemisphere seasons. When the Sun is at its northernmost point, it is high in the northern hemisphere, and it is summer there. However, since the Sun appears high in the northern hemisphere sky, it will appear low in the southern hemisphere sky, making it winter there. The reverse is true for the winter solstice. Southern hemisphere seasons are the opposite of those in the northern hemisphere.

Because of the Sun's apparent motion on the ecliptic, the stars visible at different times of the year change. When the Sun is at a given position on the ecliptic, stars and constellations near that location on the celestial sphere will *not* be seen, because they are up *with* the Sun during the day. Those stars are all *still there*, but we cannot see them because their light is washed out by the incredibly bright light of the Sun. The stars and constellations that *will* be seen are those on the opposite side of the celestial sphere.

As the Sun moves around the ecliptic throughout the year, different stars become visible at night. This is the reason for constellations being seasonal. For example, we expect to see Orion in the winter, but never in the summer, while the opposite is true of Scorpius. Capricornus may travel with the Sun in February, but it will be in the night sky in July. The night sky for a given season is opposite the Sun's position on the celestial sphere at that time of year. Stars and constellations are visible at their highest altitudes at midnight during their season.

The constellations that lie on and around the ecliptic are known as the constellations of the *zodiac*. Historically, the zodiac had been divided into twelve constellations, so the Sun spends *about* a month *in* each constellation. Among the eighty-eight constellations into which the sky has been divided in modern times, there is actually a thirteenth constellation, called *Ophiuchus*, which also lies along the path of the ecliptic, so it is also considered a member of the zodiac.

2.3 Explanations for the Motions

The observed motions of the stars and the Sun can be explained simply as they are seen from Earth, as if Earth is stationary and everything moves around it. The sky appears to rotate around Earth once per day, and the Sun appears to move *with* the sky each day. The Sun also appears to revolve around the Earth on the ecliptic once per year. This is an Earth-centered, or *geocentric*, perspective.

There is also a Sun-centered, or *heliocentric*, perspective. From this point of view, daily motion of the sky is attributed to Earth rotating, once per day, from west to east *underneath* an unmoving or stationary sky. This causes the sky to appear to move in the opposite direction each day. This is just like riding a merry-go-round. Everything seems to be moving around you in the direction opposite your motion.

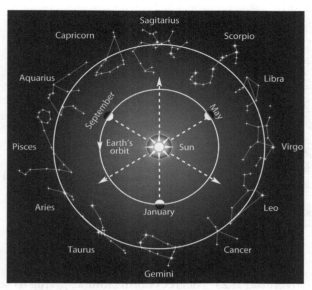

Figure 2.12 Earth's orbit around the Sun causes the Sun to appear to move through the stars and causes different stars to be visible at night throughout the year. This is why we see different groups of stars at night each season.

The Sun's annual motion can be attributed to Earth revolving in an orbit around the Sun once per year, making the Sun only *appear* to be moving along the ecliptic, causing different stars to be visible at only certain times of the year. See Figure 2.12.

From the heliocentric perspective, the seasonal changes in the Sun's daily path are explained by attributing the tilt not to the ecliptic but rather to that of the Earth's rotational axis. As Earth revolves around the Sun once per year, the Earth's axis always stays pointing in the same direction in space, toward the distant North Star, Polaris, causing different parts of Earth to receive different amounts of sunlight. This is shown in Figure 2.13.*

For those of us who live in the northern hemisphere, the angles of the Sun's rays are larger on the summer solstice. The North Pole is tilted toward them so the Sun's rays

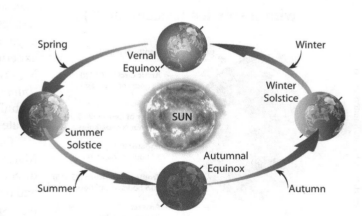

Figure 2.13 As Earth orbits the Sun, its tilted rotational axis causes the seasons (sizes and distances in the figure not to scale).

summer solstice (June 21)

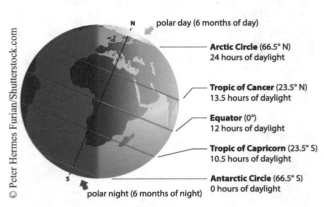

Figure 2.14

make larger angles with Earth's surface, causing the Sun to appear higher in the sky. Also note that as Earth rotates, the northern hemisphere locations will spend more time in daylight and less in darkness with the opposite being true in the southern hemisphere, where it is winter. If the Earth's axis had no tilt, we would not experience seasons. At any given location, the angle and duration of sunlight would not change. However, the 23.5-degree tilt of the Earth helps to tie together the idea of solstices and equinoxes, and also helps to explain some important locations on the Earth.

On the day of the summer solstice (around June 21), the noon sun is directly overhead at 23.5 degrees north latitude, giving us some observational support that the axis is indeed tilted 23.5 degrees. On a globe or map of the Earth, this special line of latitude is known as *the Tropic of Cancer*. No one living north of this line will ever see the Sun directly overhead at their location, but everyone in the northern hemisphere will experience the greatest duration of sunlight. In fact, anyone living north of roughly 66.5 degrees north latitude—*the Arctic Circle*—will experience at least one full day of sunlight, with the extreme being the North Pole at 90 degrees north latitude and experiencing about six months of daylight.

On the day of the winter solstice (around December 21), the noon sun is directly overhead at 23.5 degrees *south* latitude, a line of latitude known as *the Tropic of Capricorn*. While it is winter for the northern hemisphere, it will be summer for the southern hemisphere. No one living south of this line will ever see the Sun directly overhead at their location. The northern hemisphere experiences its shortest duration of sunlight, and anyone living above the Arctic Circle will experience at least one full day of no sunlight. Meanwhile, the southern hemisphere has *the Antarctic Circle* at roughly 66.5 degrees *south* latitude and the South Pole experiencing the greatest durations of sunlight.

On the days of the equinoxes (roughly March 21 and September 21), the noon sun is directly overhead at the equator. The daylight and darkness hours are roughly equal for both hemispheres, and these are the two days each year when it is appropriate to say that the Sun rises due East and sets due West.

winter solstice (December 21)

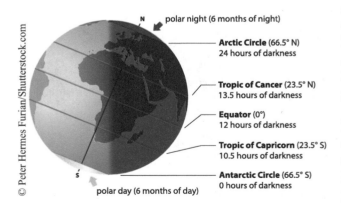

Figure 2.15

2.4 Motions of the Moon

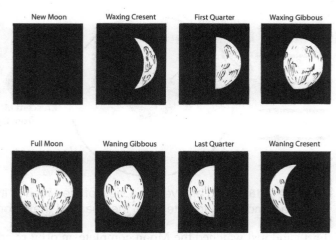

*The Moon goes through the most rapid changes in both appearance and position of any object in the sky. The variations in the amount of the illuminated portion of Moon's surface that is visible from Earth are called *phases*. See Figure 2.16.

Lunar Phases

The Moon cycles through its phases in a period of about 29.5 days. The approximately thirty-day length of our calendar *month* is based on the Moon's cycle of phases. In fact, the word "month" is

Figure 2.16 Lunar phases.

derived from the word "Moon."* One full cycle of lunar phases is known as a *synodic month*. Since a synodic month is a little shorter than most calendar months, it is possible, but not common, to get two full-moons in the same month, with the second moon called a *blue moon* and giving rise to the expression "once in a blue moon."

*From the new to full phases, the Moon grows brighter so these are called *waxing* phases. The phases from full to new, when the Moon grows dimmer, are called *waning* phases.

As the Moon orbits Earth, the Sun-facing side of each object is illuminated, while the sides opposite the Sun are dark, as seen in Figure 2.17. How the Moon will look from Earth at any given position in its orbit depends on how much of the Sun-facing side of the Moon can be seen from Earth at that time.* It comes down to geometry. Each day the angle between the Sun, Earth, and Moon changes, and we see that in the gradual change of what we perceive as phases.

Each night you see the Moon, it is a little further east than it was the night before at the same local time. In other words, each night the Moon crosses the meridian a little later—about fifty minutes later. Each day the Moon has traveled a little further in its orbit around the Earth, so it won't be in the same place in the sky twenty-four hours later.

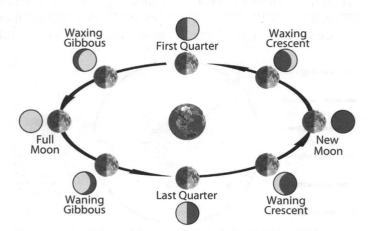

Figure 2.17 The phases of the Moon in the different positions in its orbit. The Sun's rays are coming from the right.

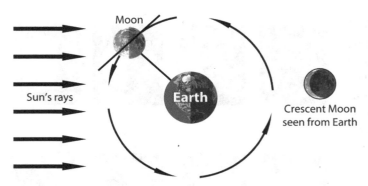

Figure 2.18 Using an overhead view of the Moon in its orbit around Earth to determine the phase of the Moon.

Recall the difference between a solar day and a sidereal day. The solar day was four minutes longer, because the Earth had to "make up" for the 1-degree it had traveled around the Sun, while that motion was meaningless relative to the much more distant stars. Something similar occurs with the Moon. Our view from Earth (synodic month) is produced by the combined motions of the Earth (as it orbits the Sun) and the Moon (as it orbits the Earth). If we ignored the Earth and just watched the Moon complete an orbit relative to the distant stars, it would be a period of just roughly 27.5 days, known as the Moon's *sidereal period*. Once again, if the Earth could magically stop revolving around the Sun while the Moon was orbiting, the sidereal period and synodic period would be the same.

*Another way to correctly visualize the phases of the Moon is with an overhead view of the Moon in its orbit around Earth, as shown in Figure 2.18. Consider yourself standing on the surface of Earth directly below the Moon's position in its orbit. Then, draw a line from your position on Earth to the Moon. Now draw another line, through the Moon, separating the side of the Moon that is facing you from the side that is not. *You can only see the side of the Moon that is facing you*. Look at the amounts of light and darkness that are facing you, and color in a circle with the correct *amount* of darkness *on the side where you see it*.

Note that the picture you are looking at in Figure 2.18 is a view from above, but you are using it to determine what you would see from your position on Earth. From the point on Earth directly below the Moon, in this example, you will see mostly darkness on the Moon's face and a little sliver of light on the left side. Coloring in a circle this way will give you a picture like the Crescent Moon shown at the right in Figure 2.18.

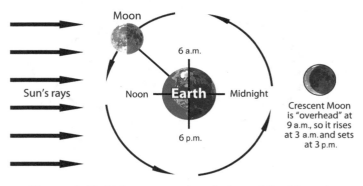

Figure 2.19 Using an overhead view of the Moon in its orbit around Earth to determine the rising and setting times for the Moon.

The local time for someone on Earth is determined by the location of the Sun in their local sky. For example, it is noon at the point on the Earth directly below the Sun. In Figures 2.18 and 2.19, that is the point on the far left of the image of Earth. At the point on Earth opposite noon, it is midnight.

Viewed from above the North Pole, Earth rotates counter-clockwise. This defines the time of day at all other points on Earth. At the point halfway between noon and midnight, at the bottom of the image of Earth in the Figure 2.19, it is 6:00 p.m. At the point halfway between midnight and noon, at the top of the image of Earth in the figure, it is 6:00 a.m. These are also the times of sunset and sunrise, the positions where an observer would cross from being in daylight to darkness and vice-versa.

The times that the Moon will be visible at any location on Earth are coordinated with its phases. The Moon will be visible from any location on Earth for the twelve hours that the Moon is above that location's local horizon. The local time on the Earth-clock when the Moon is directly above a given location will be six hours after the time at which the Moon rose and six hours before the time at which the Moon will set.

In Figure 2.19, the Moon is directly above 9:00 a.m. on the Earth-clock, so it rose six hours earlier, at 3:00 a.m. It will be at its highest altitude directly above due south at 9:00 a.m. and will set at 3:00 p.m. A similar diagram can be used for any other phase of the Moon.*

For the sake of reference, think of four lunar phases that represent specific days that you may see on a calendar, and to make things easier use four weeks rather than the actual 29.5 days of the synodic month. *New Moon* is the phase that cannot be seen. On the day of the new Moon, the Moon is between the Earth and the Sun, a condition known as *syzygy*, an alignment of three celestial objects. With this alignment, the Sun prevents us from seeing the new Moon. A new moon rises and sets with the Sun. As a ballpark idea, think of a new moon as rising in the east at 6:00 a.m., crossing the meridian at noon, and setting at 6:00 p.m.

For roughly six days after the new Moon, we have a *waxing crescent Moon.* Note that "waxing crescent" is a general description for multiple nights, while "new Moon" was a specific date on the calendar. Each night the crescent gets a little larger. At the end of one week—one quarter of the way through our four weeks—we come to the *First Quarter Moon.* We are one quarter of the way through our phases. Some people call this a half-moon because they see half of the lit face, but that could be confusing later. As an added observation, when you see a "waxing" moon, the lit portion always appears to be to your right as you see pictures or step outside to see for yourself. Don't forget where the light is coming from! The position of the setting Sun should help make sense of this.

After the day of the first-quarter moon, the waxing continues with roughly six nights referred to as *waxing gibbous.* Each night we see a little more of the lit surface, until we finally reach *full moon* half-way through our four weeks. Along with new moon and first quarter, full moon is a specific event you can find on a calendar. Generally speaking, a full moon will rise around 6:00 p.m., cross the meridian at its highest point around midnight, and then set in the west around 6:00 a.m.

After the full moon, there are roughly six nights of *waning gibbous,* which leads to a *third-quarter moon* (sometimes called the last quarter moon), followed by roughly six nights of a steadily diminishing *waning crescent*, and then it's back to another new moon. In the "waning" stages, the lit portions now appear to be on the left side as the Moon chases down a rising Sun.

© SmartS/
Shutterstock.com

Figure 2.20

Figure 2.21 The near side (left) and far side (right) of the Moon.

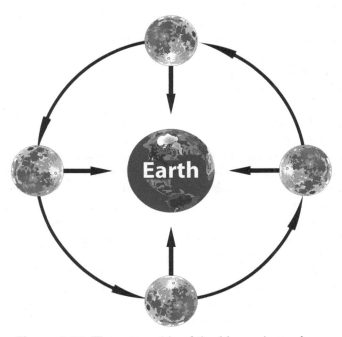

Figure 2.22 The same side of the Moon always faces Earth because the Moon rotates in the same amount of time that it takes to revolve around Earth. Note the changes in direction of the arrow, representing the near side, coming from the Moon in each position in its orbit.

The Far Side of the Moon

*As the Moon orbits Earth, the same side always faces toward Earth. So, there is a far side of the Moon that no observer on Earth ever sees. The first humans to see and photograph the far side were the astronauts orbiting the Moon in the *Apollo 8* spacecraft in late 1968.

A common misconception is that the Moon has a far side because it does not rotate. The Moon does rotate. However, the reason that the same side of the Moon always faces Earth is that the Moon's rotation period is exactly the same amount of time that it takes to revolve around Earth. The result, as shown in Figure 2.22, is that the same side of the Moon always faces Earth.

2.5 Eclipses

From Earth, the Sun and the Moon both appear to be about the same angular size, about a half-degree. A pinky-finger at arm's length is about a half-degree, so it will block out the Sun and/or the Moon. This is always true whether the Sun or the Moon is high overhead or close to the horizon. They both seem much larger when close to the horizon, since there are objects near them with which they can be compared. This can be and has been tested. No matter how large either object looks, a pinky finger can still cover them both. Try it with a full moon.

The observation that the Sun and the Moon have the same angular size (about 0.5 degrees of arc, or thirty arc minutes) is due to the fact that the Sun is 400 times larger than the Moon, but also 400 times further away. This is also the reason that events known as *eclipses* are so interesting. As indicated by the name, an *eclipse* can occur when the Moon is on or near the ecliptic. If the Moon is on the ecliptic when it is in between the Sun and Earth (in the new moon phase), it will block out the view of the Sun from Earth. This event is called a *solar eclipse*. If, on the other hand, the Earth is between the Sun and the Moon (the full moon phase), Earth's shadow will cover the Moon. This is a *lunar eclipse*.

Eclipses, however, do not occur at every new and full moon. This is because the Moon's orbit is tilted about 5 degrees with respect to the ecliptic plane. The ecliptic and the Moon's orbit are shown in geocentric perspective in Figure 2.24. The intersection of the ecliptic plane and the plane of the Moon's orbit is called the *line of nodes*. The Sun traverses the full circle of the ecliptic once per year so it will be on the line of nodes twice per year. If the Moon, in its monthly orbit around the Earth, crosses a node *when the Sun is also at a node*, there will be an eclipse. If the Sun is not at a node, there will *not* be an eclipse.

The times that the Sun is at or near a node are called *eclipse-seasons*. The Moon usually comes near enough to a node when the Sun is near one that eclipses occur about twice per year.

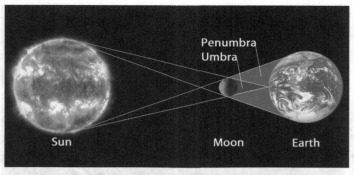
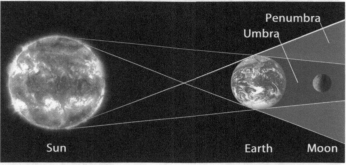

Figure 2.23 Solar (top) and lunar (bottom) eclipses.

Figure 2.24 A total solar eclipse (left) and a total lunar eclipse (right).

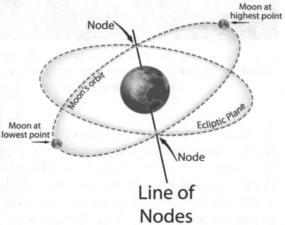

Figure 2.25 Eclipses occur when the Moon is on the line of nodes during either a full moon (lunar eclipse) or a new moon (solar eclipse).

Figure 2.26 Annular solar eclipse.

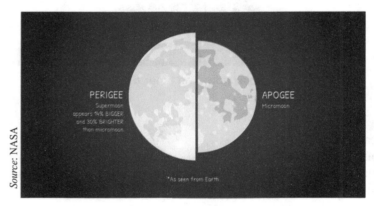

PERIGEE
Supermoon
appears 14% BIGGER
and 30% BRIGHTER
than micromoon

APOGEE
Micromoon

*As seen from Earth

Figure 2.27 It's hard to notice a difference from full moon to full moon, but this side-by-side comparison shows the two potential extremes.

If the Sun and Moon are at the same node, a solar eclipse occurs. If they are at opposite nodes, a lunar eclipse occurs. If the Sun and the Moon are not exactly lined up but are close, a partial eclipse occurs.

Referring back to Figure 2.22 will show that most people are likely to see more lunar eclipses in their lifetime than solar eclipses. During a lunar eclipse, anyone on the entire nighttime side of Earth (the side facing away from the Sun) can see the full moon eclipsed.* The Earth casts a very large shadow on the much smaller moon. The Moon passes into the darkest part of the shadow—known as the *umbra*—and the process can take several hours. During a solar eclipse, however, only people in the specific position on Earth covered by the Moon's shadow will see the solar eclipse. The Moon's umbra is relatively small, and at a given location on Earth the eclipse will last just a matter of minutes. In order to observe *totality* (the Sun appearing completely covered) an observer must be in the narrow region of the Moon's umbra. Others may observe a partial solar eclipse if they are in the part of the shadow called the *penumbra*.

There is a special type of solar eclipse called an *annular eclipse*. In an annular eclipse, the Moon passes between the Sun and the Earth in the right position, but the Moon does not cover the Sun's face entirely, leaving the edges visible. This is because of the special case where not only are the Sun, Moon, and Earth lined up correctly, but the Moon is at *apogee*. As the Moon orbits the Earth, the orbital path is slightly elliptical. Sometimes it reaches its nearest point to Earth (*perigee*), which explains the phenomenon of the "Super Moon." The Moon looks a little bigger at full moon if the full moon is also at perigee. Likewise, the new moon looks a little smaller at apogee (a "micromoon") and can't quite cover the face of the Sun during an eclipse.

*Many people in the United States witnessed a solar eclipse that followed a path from the northwest to the southeastern parts of the country on August 21, 2017. Another that will be visible from the southwest to the northeast will occur on April 8, 2024. The path of the Moon's shadow for both of these is shown in Figure 2.28.

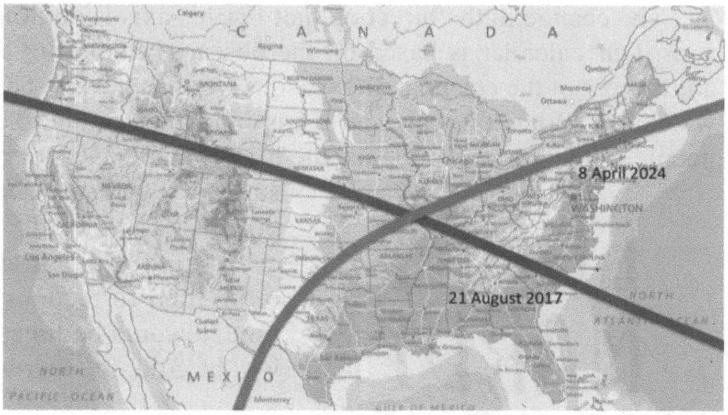

2.6 Precession

Figure 2.28 Paths of the visibility of the 2017 and 2024 total solar eclipses across the United States.

The motions discussed thus far are *rotation*, an object spinning on an axis, and *revolution*, one object in orbit around another. The rotation of Earth is responsible for the *daily motion* of the sky, while the revolution of Earth around the Sun is responsible for *annual motion* of the sky.

A third motion, not yet discussed, is *precession*. Precession is likely most familiar as the wobbling motion of a top when it is running out of energy and about to fall. The rotational axis traces out the shape of a cone, as shown in Figure 2.29. Due to precession of Earth's rotational axis, the north celestial pole, the point in the sky directly above Earth's North Pole, would trace out a circular path through the stars about once every 26,000 years.

Although it would take 26,000 years, much longer than Earth's daily rotation or its annual revolution around the Sun, the effects of the precession of Earth's rotational axis can be noticed in much smaller amounts of time. Since precession actually changes the position of the north celestial pole, the star located at or closest to this point will *not* always be the same. Polaris is our current North or *Pole Star* and obviously has been ever since it was first named. However, other stars have been the pole star in the past, and different stars will be in the future. For example, around the time of Abraham, the star Thuban from the constellation Draco was likely the Pole Star.

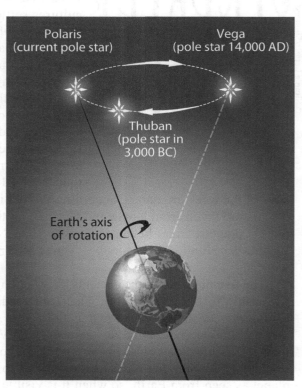

Figure 2.29 Precession.

The change in the direction that Earth's axis is pointing also affects our perception of the seasons. Our calendar is based on the observed motion of the Sun. If precession occurred without adjustment of the calendar, the seasons would appear to be occurring in months different from what we expect. When our calendar reads June, the northern hemisphere is pointed toward the Sun, so people in the north experience the high Sun angles and long daylight periods that result in warm summer weather. After about 13,000 years, half of a complete precession cycle, the northern hemisphere will be pointing away from the Sun at that particular position in the Earth's orbit. If no corrections were made to our calendar in the meantime, it would be June, but the Sun will be low in the sky and it will be cold. In other words, it would be winter in June.*

LUNDI
MARDI
MERCREDI
JEUDI
VENDREDI
SAMEDI
DIMANCHE

© Simple Letters/Shutterstock.com

Figure 2.30 The days of the week in French, starting with Monday.

2.7 Planetary Motion

Where did the seven-day week come from? We have seen that the day is based on Earth's rotation, the month is based on the cycle of lunar phases, and the year is based on Earth revolving once around the Sun. While Israel had a seven-day week tied to the biblical creation narrative, not all ancient cultures settled on a seven-day week right away, but today the names of those seven days offer a clue, especially in French or Spanish. Even in English, Sunday and Monday are pretty obvious in terms of their connection to the Sun and moon (Latin—"luna"). These are two objects that didn't behave like the thousands of stars in the sky that seemed to have fixed positions on the celestial sphere. However, there were five additional objects that appeared to wander among the stars. The Greek word for "wanderer" gives us our word *planet*. The inclusion of the planets resulted in seven objects that behaved differently from all the stars, and our seven days of the week are named accordingly. That is not as obvious in English as the names of the planets are taken from Norse and German mythology rather than the Roman names we know for the planets. Mars becomes Tuesday (the Norse god of war was Tiw), Mercury gives us Wednesday, Jupiter for Thursday, Venus for Friday, and Saturn remains easy to recognize in Saturday.

The five planets that the ancients knew—Mercury, Venus, Mars, Jupiter, and Saturn—are readily visible to observers today. They can predictably be found near the ecliptic, wandering among the constellations known as the zodiac. Most are brighter than any of the stars in the sky, and they do not "twinkle." Mercury and Venus are always relatively close to the Sun, so they are visible either close to sunset or close to sunrise, not far off the horizon. Mercury sticks so close to the Sun that it can be a challenge to locate much of the time. Venus is the brightest planet as seen from Earth, so when it is visible it is easy to identify. Mars, Jupiter, and Saturn all demonstrate a peculiar type of wandering called *retrograde motion*, in which they appear to slow down and change direction over a period of months. That motion will be explained in a later chapter, but it baffled many ancient astronomers.

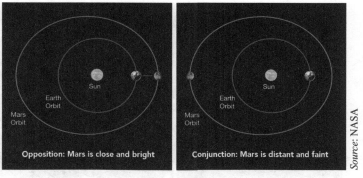

Figure 2.31

At any given time, the other planets may be in a variety of positions or *configurations* relative to the Earth. For example, picture a line running from the Sun, straight through the Earth. If any of the outer planets (Mars, Jupiter, Saturn) fall on that same line so that the Earth is directly between the Sun and the other planet(s), we say they are in *opposition*. Among other things, this means the other planet(s) will be closer to the Earth than usual and will in turn appear brighter than usual. On the other hand, if Mars, Jupiter, or Saturn were in line with Earth but on the other side of the Sun (and farthest from the Earth) we say they are in *conjunction*. In more general terms, "conjunction" can also be used to describe two objects that look very close in the sky, such as "the conjunction of Saturn and Jupiter" from 2020. For Mercury and Venus, there is no opposition since

Figure 2.32 Jupiter and Saturn in 2020.

Earth can never be between them and the Sun. When either of the planets are closest to the Earth we call it an *inferior conjunction*, and when they are furthest away and on the other side of the Sun we have a *superior conjunction*.

There will be more to say about each of the planets in our solar system in future chapters, but if you are interested in finding them in the night sky, imagine an arc across the sky where the Sun travels during the day (the ecliptic) or look for the zodiacal constellations. Then look for any unusually bright objects that don't twinkle like the neighboring stars. There are a wealth of free, online resources that will tell you exactly which planets are currently visible in the night sky and how to find them.

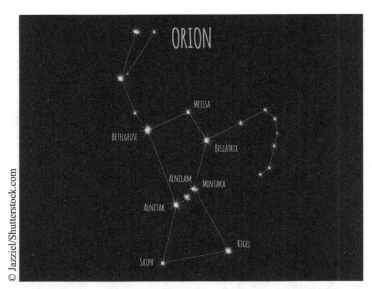

Figure 2.33 While the constellation Orion kept its Greek name, most of its bright stars have names from Arabic.

2.8 More about Stars

While each of the eighty-eight constellations has an official name, individual star names are not always as formal. Many of the very brightest stars have actual names, and these are often Arabic. During the Middle Ages, Arab astronomers translated Greek works and tended to keep the traditional constellation names while often changing star names. For example, Rigel, Betelgeuse, Aldebaran, Dubhe, Thuban, and Vega are all Arabic names. Over the years, a number of different systems emerged to name and catalog stars. In the early 1600s, Johannes Bayer introduced a method that distinguished the stars within a constellation by assigning letters from the Greek alphabet—alpha, beta, gamma, etc. In general, the brightest star would have the alpha designation, the second brightest would be beta, and so on (but Bayer was not always consistent with his own system). Sirius, the brightest star in Canis Major was named "alpha Canis Majoris." In the early 1900s, the Henry Draper Catalogue organized and identified stars based on their spectral classifications with names like HD 48915. Beyond the very brightest, well-known stars, naming systems can become confusing as any one star may have many different designations.

One simple observation that can be made on any clear night is that some stars are brighter than others. Ancient astronomers certainly knew this, and over 2,000 years ago the Greek astronomer Hipparchus produced early maps of the night sky and organized the stars by brightness. Using just naked-eye observations, Hipparchus developed a system for

Figure 2.34 Even a casual look—or a look from the International Space Station—shows that some stars are brighter than others.

star *magnitude*. The few very brightest stars he called stars of first magnitude. Second magnitude stars were a bit dimmer. His system went to sixth magnitude for the dimmest visible stars. Magnitude then is synonymous with brightness, and the larger the magnitude value, the dimmer the object.

This system is still in use today, although it has been refined and modified. For example, Sirius is assigned a magnitude value of −1.5. Venus is −4. Numbers can also be assigned to incredibly dim objects that are only visible in a telescope, with magnitude values well over 20. The principle is the same—the larger the number, the dimmer the object—but the numbers range both above and below the six magnitudes proposed by Hipparchus.

The refining process was made possible as astronomers were able to actually measure how much light a particular object was giving off. They took careful measurements and made precise calculations. The technical term for the light energy being measured is *flux*. Hipparchus was trying to determine flux with his naked-eye, so while his idea has survived, the calculations have been tweaked. There are relatively simple calculations that allow astronomers to convert between magnitude and flux. The key number to remember is 2.5 when making these calculations.

The relationship between magnitude and flux is such that for each unit of magnitude, the flux changes by a factor of roughly 2.5 (it is 2.511886, but this will make it easier). That means that a star with a magnitude of 1.0 is 2.5 times brighter than a star with a magnitude of 2.0; a star with a magnitude of 4.5 is 2.5 times brighter than a star with a magnitude of 5.5; and so on. If there is a difference in magnitude of 2, then the difference in brightness (flux ratio) is 2.5 × 2.5; if there is a magnitude difference of 3, then take 2.5 × 2.5 × 2.5 (which is roughly 16).

If the difference in magnitude is a less convenient number, like 3.3, then just take 2.5 and raise it to the 3.3 power on a calculator. Type 2.5 ^ 3.3 and you should get an answer of about 20.6. As long as you know what the difference in magnitude is between two objects, you can calculate the difference in brightness by taking 2.5 ^ (X)—where X is the difference in magnitude.

What if the question is turned around? What if the difference in brightness (flux ratio) is what is given, and the question is asking for the difference in magnitude? Don't panic. The magnitude difference can be found by taking 2.5 × (log brightness). To show how that will work, reverse a question from the previous paragraph. If the difference in brightness is known to be 20.6, what is the difference in magnitude? The difference in magnitude = 2.5 × log(20.6). We get the answer we should expect, about 3.3.

Example 1: If the magnitude difference between two stars is 5.76, what is the flux ratio (difference in brightness)?

Type 2.5 ^ 5.76

2.5 ^ 5.76 = 196 (One star is 196 times brighter than the other)

Example 2: One star is 50 times brighter than another star. What is the difference in magnitude?

Type 2.5 × log(50)

2.5 × log(50) = 4.25

2.9 Getting Outside

A good place to begin for the would-be amateur astronomer is naked-eye astronomy. The Psalmist wrote "the heavens declare the glory of God" long before the telescope. While we have all sorts of wonderful technology and resources, there is no substitute for getting outside on a clear, dark night and enjoying those same heavens for ourselves, but fewer and fewer people do.

One problem that would have been unimaginable to someone enjoying a clear night 3,000 years ago on a hillside in Palestine is light pollution. It has become increasingly difficult to find truly dark skies, and anyone born and raised near an urban area will be stunned the first time they find a view away from all of the lights. Finding a good place to star-gaze can be a challenge, but it is important for any hope of success. Even small-town residents may not realize how much their lighting compromises their view of the sky. The problem of light pollution is serious enough that there are now lists of "dark sky" locations around the United States, including a number of National Parks and State Parks.

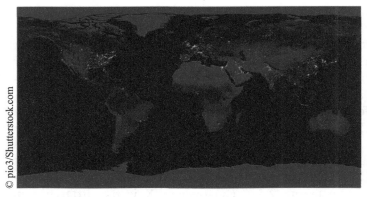

© pio3/Shutterstock.com

Figure 2.35 Light pollution is a serious problem in many parts of the world.

Besides searching out a dark spot for observing the night sky, an observer needs to have a sense of direction and position as mentioned earlier (azimuth, altitude, meridian, etc.). If you don't know which direction you are facing it can be slow going. The other helpful skill to develop is getting a sense of angular size and distance using the tricks mentioned before with an outstretched arm, hands, and fingers.

What an observer will see depends on the time and date as well as their location on the Earth. Because of the Earth's rotation and revolution, what someone will see can change with the clock and with the calendar. Depending on prior knowledge about the night sky, it can be a little overwhelming. One good tactic is to try to locate relatively easy to find "landmarks" in the night sky. Among the stars, there are several of these in the form of constellations and asterism that can give you a reference point to work from.

The best known landmark for U.S. observers is the Big Dipper. The Big Dipper is an asterism found in the constellation

© Allexxandar/Shutterstock.com

Figure 2.36 The stars of the Big Dipper.

Ursa Major. This particular landmark is circumpolar, so it is visible on any night, all year round except for brief periods if you live in the deep south. The front of the cup of the Big Dipper has two "pointer stars" that always lead to Polaris, the North Star. If you follow an imaginary line from the star at the bottom front of the cup through the star at the top front, it leads to Polaris. The Big Dipper can also help us locate seasonal objects. For example, in the spring and summer, if you follow the curve of the handle away from the cup and continue along that curve then you can "arc to Arcturus." Arcturus is an extremely bright star in the constellation Bootes. In the winter and spring, the two stars forming the rear of the cup can point you to the constellation Leo, if you extend a line from the upper rear of the cup (where the handle attaches) through the lower rear star. Follow this line across the sky until you reach what looks like a backward question mark made of stars, including bright Regulus, and you have Leo. With some practice, the Big Dipper can help an observer locate at least seven different objects of the night sky over the course of the year.

Perhaps the second best landmark is the constellation Orion, which is visible from late fall through early spring. Recall that constellations in our southern sky are all seasonal, and reflect the fact that they rise four minutes earlier each day due to the Earth's revolving around the Sun. In late November, Orion is rising in the East around 9:00 p.m., while in mid-April, Orion is setting in the West at that same time. As a point of reference, consider one of the prime times to see Orion high in the sky—February 1, around 9:00 p.m., as Orion crosses the meridian in the southern sky.

While many constellations bear little resemblance to their mythological personas, you can see how Orion creates the form of the hunter. Orion has a particularly bright, red shoulder star named Betelgeuse, while his knee on the opposite side is marked by a very bright blue star named Rigel. Only the brightest stars have colors that are discernible to the naked-eye, and with a clear, dark sky these colors are obvious. Perhaps the most famous feature, however, is Orion's belt—an asterism made of three more bright stars, though not as bright as Betelgeuse or Rigel. The belt is often what observers find first when locating Orion.

Like the Big Dipper, Orion can provide an important landmark for locating other objects. Using the February 1 reference point, angle down the line of the belt about 20 degrees toward the east and you will come to the brightest star visible from Earth—Sirius. Imagine a straight line through the two shoulder stars going east, and you will reach the bright star Procyon. Betelgeuse, Sirius, and Procyon create an asterism known as the

Figure 2.37 The constellation Orion, and the asterism known as the Winter Triangle.

© Matsumoto/Shutterstock.com

Figure 2.38 Stars of the Summer Triangle with the Milky Way. From the upper left, clockwise, Deneb, Vega, and Altair.

Figure 2.39

Winter Triangle. Besides finding some of the brightest stars in the winter sky, you have also located the constellation Canis Major (with Sirius) and Canis Minor (with Procyon). Follow the belt again, but in the opposite direction, and about 20 degrees away is the bright star Aldebaran in the constellation Taurus. Locating Taurus can also help you find a beautiful star cluster known as the Pleiades, which is one of several celestial objects mentioned in the Bible (e.g., see Job 9:9, and 38:31–32). Run an imaginary diagonal line from Rigel through Betelgeuse and about 40 degrees away you will find the "twin" stars Castor and Pollux in the constellation Gemini. As with the Big Dipper, the more you use Orion as a landmark, the more you can find.

During the summer months, an additional landmark appears high in the sky. Look straight up in late July and early August as the Sun goes down, but you can find it in less prominent positions for months. As mentioned earlier in this chapter, the Summer Triangle is made up of three very bright stars, each from a different constellation. Vega is the brightest star of the constellation Lyra, Altair is the brightest star of Aquila, and Deneb is the brightest star of Cygnus. The stars of the triangle will be among the very first to become visible as it begins to get dark on a summer night. Besides being made up of bright stars from different constellations, the Summer Triangle and the Winter Triangle share one other feature. If you have a clear, dark sky, you can see the Milky Way passing through each triangle in their respective seasons. Sadly, many locations have too much light pollution to enjoy seeing the Milky Way.

Besides stars, asterisms and constellations, there are other objects that are interesting targets for naked-eye astronomy. As previously mentioned, there are five planets that wander among the stars and are visible along the ecliptic at various times of the year. If you live in an area with dark skies, you have probably seen plenty of "shooting stars." There are predictable times when the Earth passes through a cloud of dust and debris left behind by a comet, and we observe a meteor shower like the Perseids that occur each August. Far less common, there are some comets that become bright enough to see with the naked-eye, such as Comet NEOWISE in the summer of 2020. A truly rare event to most of the United States would be to catch a glimpse of the Northern Lights (*Aurora Borealis*), although the further north you live, the greater the likelihood of getting a look. The Northern Lights are the result of varied amounts of high-energy particles from the Sun interacting with Earth's magnetic field, so unlike a meteor shower or seeing a comet, the Northern Lights are not very predictable, and many in the United States will never see them. Alaskans, enjoy the show!

With very few exceptions, the topics in this chapter have dealt with objects and events that any observer can see for themselves. Even when you *don't* know what you are looking at in the night sky, it is still glorious. God speaks through his Word, but he also speaks through his works, tainted though they may be in a fallen

Figure 2.40 Comet NEOWISE was visible for weeks in 2020.

Figure 2.41 The Northern Lights are an incredible sight, but they would be a very rare treat in Lynchburg, VA.

world. Astronomy lets us gain an even greater appreciation of those works, and for the amateur astronomer, this chapter has provided some good places to get started with no equipment necessary. If you need help getting started with making naked-eye observations, visit Appendix B for a list of free resources.

Appendix B

The Historical Roots of Modern Astronomy

Chapter 3

> *"I thank You, God, Creator, because You have blessed me with the joy of your creation; I rejoice in the works of your hands. I have used the mental talents You have given me and I have now completed the work to which You have called me. I have revealed the glory of your works to those who will read my explanations, at least that part of your infinite riches which my limited mind could grasp."*

> **—Johannes Kepler**

*THE SKY HAS FIGURED PROMINENTLY in many cultures since the beginning of civilization. The position of the Sun, Moon, and stars were often used for navigational purposes. For example, the angular distance from the North Star (Polaris) to the horizon is equal to the observer's northern latitude. This information was extremely valuable for sea voyages that traveled great distances from the European continent throughout the northern hemisphere.

The changing appearance of the sky was also linked to the change in seasons for many societies that relied on agriculture to provide food. When to plant crops and when to expect the rainy season was based on astronomical knowledge that was passed down from one generation to another.

Another important aspect of life for most cultures was the connection between the movement of objects in the sky and religious and philosophical views that people held. Seeking an explanation of what one sees in the sky is part of the natural curiosity that humans have possessed for millennia.

Due to the importance that many cultures placed on what was happening in the sky, many societies built structures

Figure 3.1 Stonehenge–Wiltshire, England.

© Justin Black, 2013. Used under license from Shutterstock, Inc.

*From *Learning Astronomy, 3/e* by Wayne A. Barkhouse and Timothy R. Young. Copyright © 2020 by Kendall Hunt Publishing Company. Reprinted by permission.

to aid in the measurement of the position of astronomical objects (e.g., Sun, Moon, and planets). The most famous of these constructions include Stonehenge in England and the pyramids of Egypt.

Although ancient civilizations did not have the extent of the scientific understanding that we have in modern times, they were able to correctly deduce several basic properties of the world around us. Ancient Chinese and Egyptian astronomers were aware of the correct length of the year more than 3,000 years ago. In ancient Greece, Aristotle and others around 350 BC had correctly inferred that the Earth was spherical in shape. Part of this knowledge was obtained from the understanding of lunar eclipses. When the Earth moves directly between the Sun and the Moon, the Earth's shadow is visible on the surface of the Moon (lunar eclipse). As the Earth's shadow first transits across the Moon or when the Moon leaves the Earth's shadow, the edge of the shadow is circular in appearance. The simplest explanation of this is that the Earth is spherical (or at least circular) in shape.

A rather ingenious effort was made by Eratosthenes from ancient Greece to measure the size of the Earth in 200 BC. Eratosthenes had heard that at noon on the first day of summer (summer solstice), objects in Syene, Egypt cast no shadows (i.e., the Sun was directly overhead). A result of this was that sunlight was able to illuminate the bottom of wells rather than strike the inside wall of the well. At the same time that the Sun was directly overhead in Syene, Eratosthenes knew that objects in Alexandria (approximately 790 km north of Syene) cast short shadows, with the Sun making an angle of about 7.2 degrees from the perpendicular (overhead) position. Eratosthenes reasoned that the difference in the angular location of the Sun was due to the fact that the Earth is spherical in shape and that by measuring the distance along the surface of the Earth between Syene and Alexandria, the circumference and radius of the Earth could be determined. Since the angular separation between Syene and Alexandria represents 1/50th of a circle (7.2 degrees/360 degrees), the multiplication of the distance between Syene and Alexandria by a factor of 50 should yield the circumference of the Earth.

Converting the unit of distance that Eratosthenes used to today's modern length measurements, the radius that Eratosthenes calculated for the Earth is 6,366 km. This value is extremely close to the accepted measurement of 6,378 km. Although there is some uncertainty in the conversion of the distance between Syene and Alexandria as measured in Ancient Greece times compared to today, the method and reasoning behind Eratosthenes's attempt to determine the size of the Earth was sound.

Long before the invention of the telescope, people made careful observations of the sky using the unaided eye. The Sun was observed to wander along a special path in the sky relative to the stars known as the ecliptic path or plane. The position of the Sun relative to the background of stars can be determined directly when the

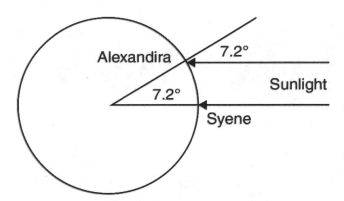

Figure 3.2 Eratosthenes' use of simple geometry, careful observations and measurements, yielded the first accurate determination of the radius of the Earth nearly 2,200 years ago.

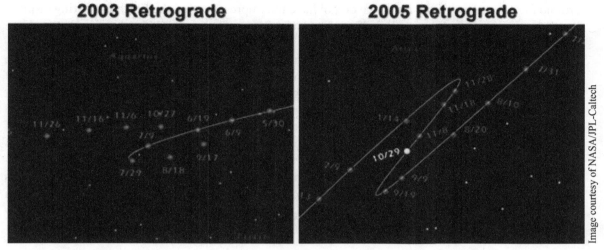

Figure 3.3 The changing positions of Mars–moving eastward (toward the left) in a prograde direction, and moving westward (toward the right) in a retrograde motion.

Moon eclipses the Sun during a solar eclipse. Ancient astronomers also realized from observations that the Sun does not change in size or brightness.

Many cultures determined that several "stars" in the night sky moved among the fixed stars. These five objects became known as the planets (Greek for "wanders") and are recognized today as Mercury, Venus, Mars, Jupiter, and Saturn. The other planets in our solar system (Uranus and Neptune) are too faint to be seen with the unaided eye. Not only did the planets move among the stars, they also changed in brightness. Planets were found to lie near the ecliptic plane, and their motion across the sky was not constant.

One particular motion of the planets that seemed out of place with the ordered nature of the heavens was the periodic backward or retrograde motion that planets displayed. In general, the planets move eastward with respect to the background stars (prograde motion), but occasionally the planets move westward for a period of time. This retrograde motion was easily observable by ancient astronomers, and begged for an explanation.

3.1 The Geocentric Model

For thousands of years, people believed that the Earth was at the center of everything (i.e., the universe). It was also believed that the Earth was at rest and that everything in the sky, Sun, Moon, planets, and stars, revolved around the Earth. In addition, it was thought that the heavens were perfect and unchanging, in stark contrast to what was happening on the Earth. This idea was quite natural in the sense that a casual glance skyward appeared to show that everything was revolving around the Earth. The Sun moves across the sky from east to west throughout the day. People observed that the Moon, stars, and planets also move toward the west during the night. Thus it wasn't hard to imagine that all astronomical objects moved around the Earth. When we walk around on the surface of the Earth we do not have the impression that the Earth is rotating or moving through space in orbit around the Sun. The lack of any sensation of motion led people to believe that the Earth was at rest.

This view of the universe, that the Earth is the center of everything, is known as the geocentric model. Both Aristotle (384 BC–322 BC) of ancient Greece and Ptolemy (AD 90–AD 168) of Alexandria, Egypt were main supporters of the geocentric view of the universe. The development of a scientific model to accurately describe nature must pass certain tests in order for the model to be adopted by others. For example, the geocentric model must be able to explain the motion of objects in the sky and accurately predict where objects will be at certain times. Part of the basis of the geocentric model was the belief that the heavens were "perfect" and unchanging. Careful observations of the movement of the planets showed that this was not the case. The geocentric model was required to explain the retrograde motion of the planets.* In geocentric astronomy, this was known as "saving the appearances." If observations did not match the model, it was up to future astronomers to figure out how to "fix" the problem without changing any of the key assumptions.

Figure 3.4 Greek philosopher Aristotle believed that Earth was the center of the universe, and that all heavenly bodies orbited the Earth in perfect spheres.

Figure 3.5 Ptolemy was a mathematician and astronomer. His geocentric model had planets moving along epicycles to explain both retrograde motion and changing brightness.

*Claudius Ptolemy in AD 140 introduced a modification of the geocentric model that included epicycles. As depicted in Figure 3.6, the planets do not trace out a path around the Earth on a perfect circle, but instead move around a smaller circle (epicycle) whose center moves around the larger circle (the *deferent*). The introduction of epicycles was motivated by the desire to explain both retrograde motion and the changing brightness of the planets. As a planet travels along the epicycle in the same direction as the larger circle, the planet will be seen from Earth as moving in a prograde direction. When the planet passes along the arc of the epicycle that is closest to the Earth, the planets appeared to move in a backward or retrograde direction. As the planet continues to move along its epicycle, it would appear from the Earth to move eastward and then westward for a period of time. The use of the epicycle thus explained the perplexing backward motion of the planets relative to the stars.

The epicycles also explained the changing brightness of the planets. As a planet moves along its epicycle, the distance between the planet and the Earth change. When the planet is furthest from the Earth it will appear dimmer than when the planet is closer to the Earth. The motion of the planet along its epicycle explains both its retrograde motion and the brightness variation.

Ptolemy published his version of the geocentric model in his book, the *Almagest*. After the introduction of epicycles by Ptolemy to "save the appearances," the geocentric model was revised several times as astronomers tried to improve the match between the predicted location of the planets and their observed positions. In a real sense, the geocentric model became complex with the use of epicycles within epicycles as errors in the position of objects in the sky were reduced to a point where the eye was not able to discriminate between the predicted and actual location of a planet.

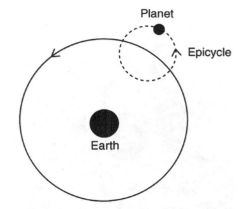

Figure 3.6 Ptolemy's introduction of epicycles helped to ensure that the predicted location of the planets matched the observations that were made at that time.

3.2 The Heliocentric Model

Approximately 2,300 years ago, the Greek astronomer Aristarchus of Samos (310 BC–250 BC) proposed that the Sun (in Greek, *"helios"*) was the center of the universe and not the Earth. He suggested that all the planets, including the Earth, revolved around the Sun, and that the Earth also rotated on its axis. The rotation of the Earth explains the observed fact that astronomical objects move from east to west during the course of the day and night. This heliocentric model was not accepted at the time since it not only went against widely held beliefs that were based on religious and philosophical arguments,* but it also lacked any compelling empirical evidence. Then, as now, people were not likely to abandon a model that appeared to work, and for that time geocentrism did seem to work. The reluctance to entertain a competing model is also understandable if the new model cannot demonstrate some superiority in predicting and explaining, and for that time heliocentrism could not.

Image © Lefteris Papaulakis, 2013. Used under license from Shutterstock, Inc.

Figure 3.7 In 1980, Greece issued a postage stamp commemorating 2,300 years since the birth of the ancient Greek astronomer Aristarchus of Samos.

Nikolaus Kopernikus.

Figure 3.8 Nicolaus Copernicus reintroduced the idea of a Sun-centered universe, with the Earth and planets orbiting the Sun in perfect circles.

3.3 Nicolaus Copernicus

*The Polish astronomer Nicolaus Copernicus (1473–1543) revitalized the heliocentric view of the universe with the publication of his book *De revolutionibus orbium coelestium* (*On the Revolutions of the Heavenly Spheres*) shortly before his death in 1543. Copernicus believed that the heliocentric model provided a more natural explanation of the universe than the geocentric model with its complex arrangements of epicycles within epicycles.* While Copernicus wrote about a radical idea, he was no rebel. Rather than being a renegade taking on the Church of his day, he was a devout, religious man who wrote "To know the mighty works of God, to comprehend His wisdom and majesty and power; to appreciate, in degree, the wonderful workings of His laws, surely all this must be a pleasing and acceptable mode of worship to the Most High, to whom ignorance cannot be more grateful than knowledge." Challenging the Ptolemaic model was in no way an attempt to challenge faith in God. Instead, astronomy for Copernicus was an "acceptable mode of worship." However, he was nervous about publishing his work and how it might be received, so he put it off until he was approaching the end of his life.

*With the heliocentric model Copernicus was able to explain that the smooth motion and unchanging appearance of the Sun was due to the fact that the Earth is orbiting around the Sun in a perfect circle at a constant speed. The changing brightness of the planets is a result of the changing distance between the planets and the Earth. Since all planets, including the Earth, orbit the Sun, the distance between planets will constantly change. The light from an object gets dimmer as the distance to the object increases (light decreases as the inverse-square of the distance),

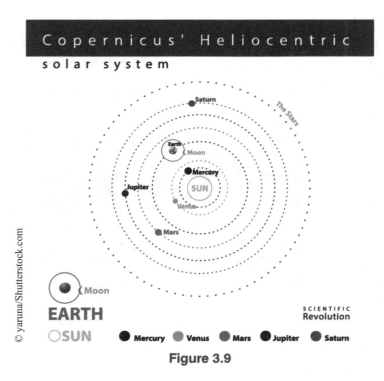

Figure 3.9

and hence the changing distance naturally explains the variation in the brightness of the planets.

A powerful aspect of the heliocentric model is its ability to explain the observed retrograde motion of planets without resorting to an ad hoc add-on to the model. Copernicus reasoned that planets closer to the Sun will take less time to complete an orbit than those further away (orbits with a smaller radius have a smaller circumference). For the planets visible to the unaided eye that are further from the Sun than the Earth, namely Mars, Jupiter, and Saturn, these planets will appear to undergo retrograde motion whenever the Earth passes those planets on the inside (Figure 3.10). This is similar to the effect when you pass another vehicle on the highway that is also traveling in the same direction as yourself. As you pass the slower vehicle it appears from your perspective that the other object is moving backward (retrograde motion). However, you instinctively know that you and the other vehicle are both traveling in the same direction.

For planets that are closer to the Sun than the Earth (Mercury and Venus), as viewed from the Earth the planets will appear on one side of the Sun and then after a certain period of time they will be visible on the other side of the Sun (early evening and early morning). This back and forth movement is the retrograde (westward) and prograde (eastward) motion.

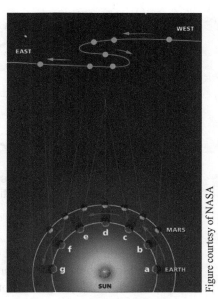

Figure 3.10 Illustration of retrograde motion of Mars as viewed from Earth. The apparent backward motion of the outer planet occurs when the inner planet is passing the outer planet.

Although Copernicus was able to explain the movement of planets in the sky using a less contrived and complicated model, the predicted position of the planets using the heliocentric model was no more accurate than those given by the geocentric view. The main reason for this is that Copernicus maintained some of the ideas of the Greek astronomers, and assumed that the planets, including the Earth, orbited the Sun in perfect circles. His model also continued to make use of some epicycles. As we will see, these erroneous assumptions prevented the heliocentric model from displacing the geocentric model from popular belief right away.

An even more troublesome aspect of the heliocentric model was the implication that if the Earth moves around the Sun, the stars should display an annual parallax. This parallactic motion is the result of viewing a star from two different vantage points. An analogue is observing your finger extended at arms' length using one eye, and noting where the finger is located relative to the background. When you switch to the other eye, the finger will appear to shift with respect to the background. Parallax is the apparent motion of some distant object that is actually caused by the motion of the observer. When a star is viewed from one side of the Earth's orbit (say in July) compared to six months later (January) when the Earth is on the opposite side of its orbit, the star should shift position relative to the backdrop of more distant stars. This effect was looked for and not seen in the time before the invention of the telescope. This lack of stellar parallax went against the acceptance of the heliocentric model (stellar parallax was first observed in 1838 by Friedrich Bessel).

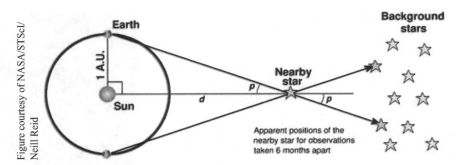

Figure 3.11 Stellar parallax is the shift of the apparent position of a nearby star relative to background stars when viewed from two different locations of the Earth's orbit.

The explanation for this apparent contradiction is that stars are so far away that the shift in position is too small to be observed without a good telescope.*

The work of Copernicus alone was not sufficient to prove heliocentrism. What Copernicus managed to do was reintroduce people to a revolutionary idea and generate further interest. The heliocentric system as explained by Copernicus was still very complex and difficult when it came to the details. It was at best marginally better than geocentrism when it came to making predictions and explaining observations, although retrograde motion did disappear. The real success of the "Copernican model" would come later, through the work of other astronomers who were inspired by *De Revolutionibus*.

3.4 Tycho Brahe and Johannes Kepler

Tycho Brahe (1546–1601) and Johannes Kepler (1571–1630) will be forever linked through history as the observer and theorist who solved the orbits of the planets. Tycho Brahe was a Danish nobleman who became interested in making accurate astronomical observations of the position of objects in the sky. Like various one-name celebrities today, we often refer to Tycho Brahe simply as "Tycho," and in his day he was as much of a celebrity as an astronomer could be. He was quirky and colorful and a bit self-important. In 1566, he got into a sword-fight with his cousin over a math equation and had his nose cut off. He had metal prosthetic noses made that he could attach to his face. *When Tycho first became interested in astronomy, he

Figure 3.12

discovered that the best astronomical tables contained many errors in the position of stars and planets. He decided to spend his time and resources to catalog as precisely and accurately as possible the position of objects in the sky.*

In 1572, Tycho made careful observations of what at the time was called a *stella nova* (a new star). Today we know it was a supernova, and we have detailed images of its remnant. From previous observations of comets, Tycho ruled that out as a possibility. He could not detect any parallax, which suggested that this "new" object was very far away. His observations undermined the claims of Ptolemy and Aristotle. Ptolemy taught that any changes had to occur relatively close to the Earth, inside the sphere of the Moon. However, the lack of parallax meant that was not the case. Aristotle taught that the heavens were perfect and immutable

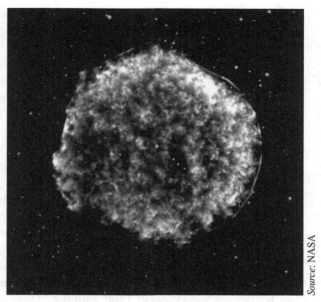

Source: NASA

Figure 3.13 Using various telescopes, we can still observe the remnant of Tycho's *stella nova*.

(unchanging), but here was clear evidence of change. Tycho published his work a year later in a book he titled *De Stella Nova*, and people took notice.

Tycho was fortunate to obtain financial resources from the King of Denmark which allowed him to build an observatory on an island in Denmark. This was before the invention of the telescope, thus the instruments that Tycho built were used to measure positions of astronomical objects using the unaided eye. The facilities Tycho built on the island represented the most state-of-the-art research labs of the day. His observatory included huge, expensive instruments and an army of staff scientists working for him. Even Tycho's sister Sophie played an important role in the work, which was rare for women during this time period. Tycho and his staff produced some of the most precise naked-eye observations of the night sky ever. By the end of his career, Tycho was charting objects to within two minutes of arc of their true positions,

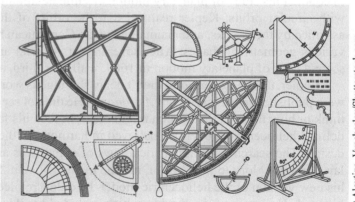

© Alevtina_Vyacheslav/Shutterstock.com

Figure 3.14 Tycho took devices for navigation and astronomy and had them built on a much larger scale to make his measurements.

just 1/30th of a degree. It would be years before anyone improved on this even after the invention and use of the telescope. Sadly, when the King died, Tycho was forced to leave Denmark and relocate in Prague.

One of the greatest achievements that Tycho was able to obtain was the decades-long record keeping of many objects in the night sky, including the planet Mars. Not only did Tycho and his workers measure astronomical positions with great accuracy and precision for his time, they also understood the uncertainty in their measurements. Thus Tycho was able to assemble the largest and most accurate collection of astronomical positions that had ever been acquired. While Tycho's observations were unmatched in both quality and quantity, his own attempts to devise a model or theory to replace geocentrism failed miserably. He did not accept the Ptolemaic model, but he also never became convinced that the Earth was moving.

*Johannes Kepler, a German mathematician during the time of Tycho, was interested in understanding the motion of the planets. Kepler believed in the heliocentric model of the solar system and wanted to determine precisely the orbital paths of the planets as they traveled around the Sun. In order to mathematically analyze the motion of the planets and thus figure out the shapes of their orbits, Kepler needed precise measurements of planetary positions with accurate estimations of their uncertainties. Kepler decided to contact Tycho Brahe and made arrangements to visit him.

Kepler visited Tycho in 1600 with the idea of going to work analyzing the tables of astronomical observations that Tycho had so carefully constructed over several decades. The story that has been told indicates that Tycho was reluctant to turn over all of his prized records to Kepler, which caused great consternation between the two gentlemen.

In 1601, Tycho attended a banquet in Prague in which royalty was in attendance. During the dinner, Tycho was in need of using the washroom but was not able to since it was bad manners to leave the dinner table while royalty were seated. Eventually Tycho was able to make it to the washroom, but he contracted a bladder infection and died eleven days later (October 24, 1601).

After Tycho's death, Kepler gained full access to all of the astronomical records that he required to conduct his study of planetary motions. While he initially tried to make Tycho's observations fit with circular orbits, Kepler realized that the orbits of the planets are not circular as had been assumed by the Greeks, and maintained in the Copernican model, and he tried to fit the orbits using various geometrical shapes. After several years of trial and error, Kepler published his first and second laws of planetary motion in 1609, while the third, and final, law was published in 1619.*

Kepler, unlike Copernicus, didn't delay sharing his work. His first two laws of planetary motion were published in his book, *New Astronomy*. He did not seem concerned about what his peers might think or how those in the Church might react. In fact, his faith helped Kepler move toward a better defined heliocentric model. He believed in harmony in the universe, put there according to God's design. The search for mathematical laws of nature was an act of worship, and discovering those laws brought glory to the Creator, not the mathematician. So when Kepler successfully developed his new version of the heliocentric model, he didn't rejoice at the triumph of science over faith. In his book *Harmony of the Worlds*, which included his third law of planetary motion, Kepler said, "Great is God our Lord, great is His power and there is no end to His wisdom." He was not shy about mixing scientific discovery and Christian faith. As a young man, Kepler had intended to become a theologian, but instead he found that he could glorify God in astronomy.

*Kepler discovered that the planets revolved around the Sun on elliptical orbits. This corrected the major flaw in the Copernican model since the supporters of the heliocentric view believed that the planets orbited the Sun in a perfect circle. Due to this modification, the position of the planets agreed more closely with observations than the geocentric model with its complicated systems of epicycles.

An ellipse is shaped like a squashed circle and defined by the length of its major and minor axes (Figure 3.15). One-half of the major axis is known as the semi-major axis, and one-half of the minor axis is the semi-minor axis. The degree of elongation of an ellipse is known as the eccentricity (ε), and is defined as the ratio of the distance between the focal points and the length of the major axis (Figure 3.16). When the focal points coincide, $\varepsilon = 0$ (i.e., a circle). The other extreme case is when the distance between the focal points approaches infinity. In this case the eccentricity approaches one ($\varepsilon \rightarrow 1$) since both the major axis and the distance between focal points is the same (i.e., a line).

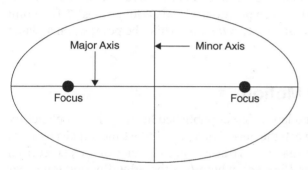

Figure 3.15 The shape of an ellipse is defined by the length of its major and minor axes. A circle is a special case where the major and minor axes are equal.

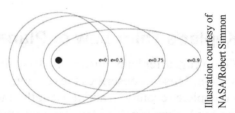

Illustration courtesy of NASA/Robert Simmon

Figure 3.16 A circle is a special case where $\varepsilon = 0$. As separation between focal points increases, the ellipse becomes more eccentric or elongated.

Kepler's First Law of Planetary Motion

With an understanding of the basic properties of ellipses, we are able to appreciate Kepler's three laws of planetary motion. Kepler's first law states that a planet travels around the Sun in an elliptical orbit with the Sun at one focus. The question that naturally arises is if the Sun is at one focal point, what is at the other? For planetary orbits in our solar system, the other focus point is unoccupied.

Although Kepler used Tycho Brahe's astronomical tables of planetary positions to determine that an ellipse fits the orbital path of the planets, he was not able to understand why this is the case. Kepler hypothesized that there may be a force of attraction between the planets and the Sun that keeps the planets in their orbits. Although Kepler did not know what this force was, he did suggest that it may be magnetism based on the work of William Gilbert. Isaac Newton solved this problem in 1687 with his publication of the universal law of gravity.

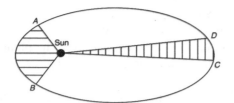

Figure 3.17 Kepler's second law states that as a planet orbits the Sun, it sweeps out equal area in an equal time interval.

Kepler's Second Law of Planetary Motion

Together with the first law, Kepler also published his second law of planetary motion which states that an imaginary line between a planet and the Sun will sweep out an equal area in an equal time interval. As shown in Figure 3.17, a planet orbiting the Sun will take the same time to travel from position *A to B* as it takes to go from *C to D. The areas enclosed between positions A and B and between C and D are equal.*

The conclusion of Kepler's second law is that planets travel faster when near the Sun than when they are further away. In order for the swept out areas of the two regions to be the same, given that the length of the sides are very different, the planet must move faster between positions *A* and *B* (sides are short) than when the planet is moving between positions *C* and *D* (long sides).* The point in a planet's orbit when it is closest to the Sun is called its *perihelion*, while the point when a planet is furthest from the Sun is its *aphelion*.

Kepler's Third Law of Planetary Motion

*Ten years after he published his first and second laws, Kepler published his third law of planetary motion. This law describes the relationship between the period of a planet (the time it takes to make a complete revolution) and the length of its semi-major axis (the average distance of the planet from the Sun). Kepler's third law can be expressed as $P^2 = ka^3$, where P is the orbital period (one year for the Earth), and a is the semi-major axis distance (1AU for the Earth). The constant k can be determined by applying the third law to the Earth. If the average distance between the Earth and the Sun is given in astronomical units (1AU), and the orbital period is in years (1 year for the Earth), then we have $k = \dfrac{P^2}{a^3} = \dfrac{(1 \text{ year})^2}{(1 \text{ AU})^3} = 1 \dfrac{\text{year}^2}{\text{AU}^3}$. If we measure the orbital period in Earth years and the semi-major axis distance in AU, we get $k = 1$ and thus we can simply use the equation $P^2 = a^3$. The consequence of Kepler's third law is that planets far from the Sun have longer periods (for Neptune, $P = 165$ years) than planets that are closer to the Sun (Mercury has an orbital period of 0.2 years).

As an example, we can use Kepler's third law to determine the orbital period of Mars based on the fact that the average distance between Mars and the Sun is 1.52 AU. We get $P^2 = a^3 \rightarrow P = a^{3/2} = (1.52 \text{ AU})^{3/2} = 1.87$ years.

Conversely, if we know that an asteroid in orbit around the Sun takes ten years to complete one orbit, the asteroid is an average of $a = P^{2/3} = (10 \text{ years})^{2/3} = 4.6$ AU from the Sun.

Kepler's three laws of planetary motion accurately describe how planets move, but do not explain why. Sir Isaac Newton took up this challenge.

3.5 Galileo Galilei

Galileo Galilei (1564–1642) is credited as the first person to build a telescope for making astronomical observations and publishing what he observed. He did not, however, "invent" the telescope. Hans Lippershey from the Netherlands is usually credited as the first person to build a telescope in 1608, although several others at that time were involved in making optical devices. In 1609, Galileo built his own telescope following the description of the instrument that Lippershey had constructed. Galileo used his device to observe the heavens and revolutionized astronomy.

With his telescope Galileo was able to observe that heavenly bodies were not perfect as claimed by the geocentric believers. Galileo observed that the Moon's surface was not smooth but was covered in craters, valleys, and mountains. The Sun, when viewed by projecting the light on a flat background, was found to have dark spots (solar magnetic storms known as sunspots). Careful observations by Galileo indicated that over the course of a month, some of the spots moved behind the Sun and back around to the opposite side. Galileo correctly interpreted the time that it takes a sunspot to return to its starting position as the rotation period of the Sun.

Figure 3.18 Galileo was the first person to use a telescope to view the heavens and publish his observations.

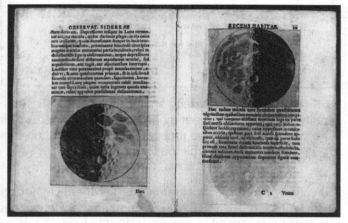

Figure 3.19 Galileo observed that the Moon had mountains, valleys, and craters—its surface was not smooth and "perfect." This figure shows drawings of the Moon that Galileo made while using his telescope.

Although telescopic observations of the Moon and Sun demonstrated that celestial bodies are not "perfect," the evidence for this had been around for quite awhile. When one looks at the Moon, for example, you can easily see that the surface is not perfect but consists of alternating patterns of dark and bright regions. There had been numerous reports that dark spots are occasionally visible on the surface of the Sun. During times when the cloud cover is of sufficient thickness, one can stare directly at the Sun without being overwhelmed by its brightness (never attempt this!). If a group of sunspots happen to be large enough at the same time, they can be observed by the unaided eye. Early reports of spots on the Sun were dismissed as being due to objects in the Earth's atmosphere. Strict believers of the geocentric Aristotelian/Ptolemaic universe refused to accept that the heavens were anything but perfect.

One of the major discoveries that Galileo made was that the planet Jupiter has a system of four moons revolving around it. Galileo first sighted these satellites in late 1609 and followed them carefully for many months (Figure 3.20) before fully realizing that these objects were in orbit around Jupiter and not the Earth. The importance of this discovery was that the geocentric model was based on the idea that everything in the heavens orbited the Earth. With a telescope, anyone could see that there were other objects in the sky that went around Jupiter and not the Earth.* Today we call the four largest moons of Jupiter the *Galilean moons*. Galileo published his discovery about the moons of Jupiter along with a variety of other telescopic observations in a small book called *Sidereus Nuncius* (the Starry Messenger). Other publications would follow, but *Sidereus Nuncius* helped make Galileo famous and advance his career.

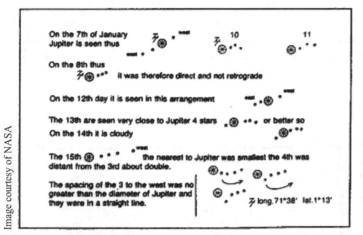

Image courtesy of NASA

Figure 3.20 This drawing shows the observations of Jupiter and its four largest moons by Galileo in 1610. This discovery demonstrated that not all astronomical objects orbit the Earth.

*When Galileo observed Venus he discovered that the planet displayed a series of phases similar to the Moon. He noticed that when Venus was at crescent phase, it had the largest angular size compared to when Venus had a different phase. Galileo also observed that when Venus was at "full" phase, the planet had the smallest apparent size through a telescope.

The discovery that Venus had phases, and that there was a correlation or relationship between the size of Venus and its phase, turned out to be the proverbial straw that broke the camel's back for the geocentric supporters. Here were new observations that had not been known when both the geocentric and heliocentric models were first constructed. One of the basic ways to test a scientific theory is to make predictions and see if they match reality. Another way is to make a new observation and see if the theory is able to explain what is observed. The geocentric model is not able to explain the relationship between the phases of Venus and its apparent size, but the heliocentric model is able to provide a simple explanation without any modifications.

Figure 3.21 clearly demonstrates that the geocentric model is not able to provide a logical explanation for the observations of Venus made by Galileo with his telescope. Although Venus will display a series of phases as viewed from the Earth, there is no geometrical arrangement of the Sun, Venus, and Earth that would allow Earth-based observers to see Venus fully illuminated (full phase). In contrast, the heliocentric model provides a natural explanation of the relationship between the phases of Venus and its apparent size as viewed through a telescope (Figure 3.22). When Venus is near the Earth (and thus large in appearance) it will display a thin crescent phase. When Venus is on the opposite side of the Sun, and thus the furthest away from the Earth, its Earth-facing hemisphere will be fully illuminated due to reflected sunlight (full phase). Since the orbital paths of the planets are not all contained in the same plane, when Venus is on the opposite side of the Sun it is usually visible above or below the disk of the Sun. Not only does the heliocentric model explain the changing phases of Venus, but it also explains the observed fact that Venus is smaller in size at full phase compared to crescent phase.

The key to resolving the debate between the geocentric and heliocentric models was the scientific testing of these ideas with new observations.* Galileo also introduced the early ideas of relativity (relative motion) and inertia to explain why we do not "feel" the motion of the Earth.

Many of Galileo's experiments were thought experiments that he would detail in his writings, rather than actual, physical experiments. For example, when he claimed that a hammer and a feather would fall to Earth at the same rate (in the absence

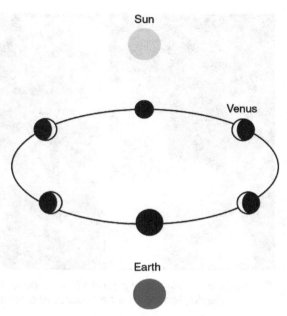

Figure 3.21 This drawing illustrates the attempt to use the geocentric model to explain the phases of Venus and their correlation with the angular size of the planet as viewed through a telescope.

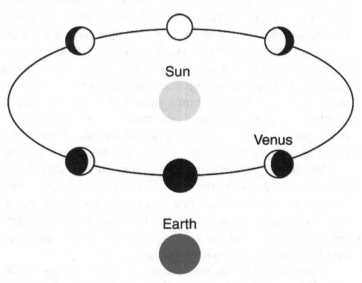

Figure 3.22 The heliocentric model offers a natural explanation of the phases of Venus and its apparent size.

Source: NASA

Figure 3.23 This grainy image shows Apollo 15 astronaut David Scott with a hammer and a feather.

of air), he could not do the physical experiment, but he did write a compelling explanation. In the case of the hammer and the feather, astronaut David Scott successfully conducted the actual experiment on the Moon, during the Apollo 15 mission in 1971.

Galileo enjoyed a certain degree of popularity and fame during his life, but by 1632 that all changed. The oversimplified, decontextualized version of events would claim that Galileo was an innocent, objective, clear-minded scientist who was attacked by religious leaders who hated science. This positivist tale claims that Galileo was arrested and threatened with death for believing that the Earth moves around the Sun.

Sadly, this is now a common misconception, despite a wealth of historical information that contradicts the claim. A similar false story appears in many astronomy textbooks that claim some years earlier, the "scientist" Giordano Bruno was burned at the stake for believing the Copernican view. While Bruno was burned at the stake (and it is never right to burn someone at the stake), it was for being a Catholic priest during the time of the Inquisition who became a theological heretic and tried to start his own cult. The ties to the Copernican view were at best incidental, but the story helps feed the false-narrative that science and religion have always been at war.

What really happened to Galileo is much more complicated, and has little if anything to do with the interplay of faith and science. While Galileo was a genius, he was also arrogant and pugnacious. He enjoyed arguing, he enjoyed humiliating his detractors, and he enjoyed being right. While his many successes brought him a period of fame, his tactless personality made him many enemies. Galileo likely overestimated how much security his fame would bring him and underestimated how many people he was offending and alienating. It may be that for some of Galileo's enemies, the Copernican view was a convenient topic to use as a means of retaliation.

By 1616, enough controversy was stirring that the promotion of the Copernican view was called into question by the Catholic Church, which of course meant that Galileo would be involved. The use of *De Revolutionibus* was suspended that year, and Galileo was called to Rome where he was given instructions to restrict his teaching of the Copernican model in hopes of quelling the controversy. It is unclear how rigid those restrictions were, but for a time Galileo shifted his attention from astronomy to other topics.

Eventually, Galileo wrote another book, *"Dialogo"* (*Dialogue Concerning the Two Chief World Systems*). As the title suggests, the book was written as a dialogue or conversation between three fictional friends. One character was Salviati (literally "the savior"), who was wise and quick-witted as he defended what was obviously Galileo's perspectives. Segredo served as a neutral, somewhat uninformed friend. Simplicio ("the simpleton") was then the friend who haplessly tried to defend Ptolemaic astronomy and other Greek ideas that Galileo obviously opposed. Whether purposeful or not, Galileo wrote in such a way that Simplicio used some of the exact arguments that were coming from the Pope himself at that time, and Galileo's detractors made sure the Pope knew it.

So to put this in context, Galileo created controversy, in part with his abrasive personality. He was warned and restricted by the Catholic Church, and his eventual response was to mock the Pope in a book that was available to anyone who was interested, with all of this occurring during the time of the Inquisition. Galileo was summoned to the Inquisition late in 1632, and in 1633 he was found guilty of disobeying the orders from 1616. He avoided torture and death, and instead was put under house arrest for his remaining years. Through it all, Galileo did not renounce or question his faith in God. His trial was not based on science versus religion. To restate the obvious—it was complicated.

Figure 3.24 This illustration from *Dialogo* shows Salviati, Segredo, and Simplicio in conversation.

3.6 Isaac Newton

The mystery behind the force that holds the planets in orbit around the Sun was solved by Sir Isaac Newton (1643–1727). While Newton is famous for his scientific genius, nevertheless he was also devoutly religious and socially awkward. By all accounts, he had an unhappy childhood and became a reclusive introvert as an adult. He would immerse himself in his work, but he had few friends and never married. He wrote extensively about theology although many of his views were unorthodox. He invented calculus, the reflector telescope, and wrote about the nature of light.

In 1687, Newton published his book *Philosophiae Naturalis Principia Mathematica* (commonly referred to as the "Principia"), which outlined his three laws of motion and the universal law of gravity. In his *Principia,* Newton also wrote: "This most beautiful system of the Sun, planets, and comets, could only proceed from the counsel and dominion of an intelligent and powerful Being … This Being governs all

Figure 3.25 Newton revolutionized physics with his laws of motion and universal law of gravity.

things, not as the soul of the world, but as Lord over all. . . . The Supreme God is a Being eternal, infinite, absolutely perfect . . . and from his true dominion it follows that the true God is a living, intelligent, and powerful Being." Newton was very transparent about worshiping God as the Creator, and that included careful study of the natural world. It is a sad irony that today the "Newtonian worldview" attempts to divorce faith from reason. Isaac Newton saw no conflict between the two.

Three Laws of Motion

*Newton's three laws of motion described the action of forces on objects and how they move in response. This began the branch of physics known as classical mechanics or Newtonian mechanics. Newton's three laws of motion can be stated as follows:

1.   An object at rest will remain at rest, or an object in motion will continue in a straight line at a constant speed, unless acted upon by a net external force. This is often called the law of inertia. *Inertia* is a property of matter that describes its resistance to change.
2.   The acceleration (*a*) of an object is directly proportional to the net force (*F*) acting on the object, and inversely proportional to its mass (*m*). We can summarize the second law as $a = \dfrac{F}{m}$ or $F = ma$.
3.   For every action there is an equal and opposite reaction.

   An example of Newton's first law of motion is the placement of an object on a table and the observed fact that the object will remain stationary unless some type of action or force disturbs the object (e.g., someone picks up the object, or the table is moved). The action of someone picking up the object means that an external force (external to the object) is acting upon that body. The more massive the object, the greater its inertia.

   An example of the second part of the first law is the motion of a hockey puck along the surface of ice. The first law predicts that the puck should continue in a straight line at a constant speed unless an external force acts on the puck. Experience shows that after hitting a hockey puck on the surface of a large lake free of snow, the puck will eventually slow down and come to a stop (assuming you are far from the edge of the lake). Does this violate Newton's first law of motion?

   Newton would answer this question by stating that since the motion of the puck is changing (i.e., the puck is slowing down or

Figure 3.26 Newton's first law of motion can be demonstrated by watching the motion of a hockey puck across the surface of ice.

decelerating), there must be an external force acting on the puck. The external force in this case is the frictional force between the surface of the ice and the bottom of the puck. You can demonstrate this by smoothing the surface of the ice, thus reducing the frictional force, and repeating the experiment to see how far the puck will travel before coming to a stop. We find that as the surface becomes smoother and smoother, the puck will travel further before slowing down. We can extend this thought experiment until the ice becomes completely smooth with no friction between the puck and the ice

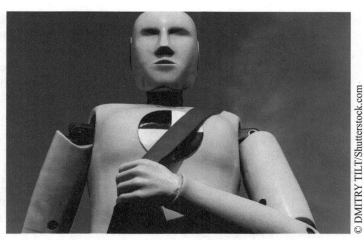

Figure 3.27 Newton's first law of motion can explain why seatbelts are so important.

surface. At this point the puck would travel in a straight line at a constant speed, thus verifying Newton's first law of motion. Of course, we are neglecting other external forces on the puck such as air resistance, etc.*

Newton's first law explains why you should always wear your seatbelt! As you travel along in a vehicle, you are "an object in motion." With a sudden stop (such as hitting a tree), the car may stop abruptly, but your inertia will cause you to continue to travel unless another force (the restraint of the seatbelt) safely halts your motion. Without a seatbelt, other forces will bring you to a stop, but with potentially deadly consequences.

*Newton's second law of motion is really a restatement in mathematical form of the first law. The equation $F = ma$ is as famous in Newtonian mechanics as $E = mc^2$ is in Einstein's relativity. The second law simply reflects the fact that a change in velocity of a body (acceleration) is proportional to the net force that acts on the body. If an object is motionless on a table, it will continue to remain motionless unless an external force accelerates the object, causing it to move. This is just the first part of Newton's first law of motion. If an object is traveling in a straight line at a constant speed (not accelerating) it will continue to do so unless a net force is applied to the object that causes it to accelerate (i.e., change direction and/or speed). The greater the force, the greater the change. This was the case for the hockey puck example where the frictional force on the puck caused it to slow down (negative acceleration or deceleration).

Newton's third law of motion, for every action there is an equal and opposite reaction, can be illustrated with the example of someone pushing hard against a wall while standing on a chair with wheels. Instinctively you know that if the person pushes hard enough against the wall, the chair will move away from the wall and the person may fall. The third law of motion states that when we apply a force on the wall by pushing on it, there will be an equal force in the opposite direction on us. This reaction force causes the chair to roll away from the wall. The wall and chair are called an action/reaction pair.

Figure 3.28 Newton's third law of motion is demonstrated in the launch of a rocket. The expulsion of hot gas from one end of the rocket allows it to move in the opposite direction.

Another example of such an action/reaction pair is the principle behind the working of a rocket. As hot gas is expelled from the end of a rocket engine, there is a reaction force in the opposite direction on the inside surface of the rocket engine nozzle. Newton's third law of motion thus explains how a rocket is able to propel itself forward, even in the vacuum of space. We see this same principle demonstrated when you let go of a balloon filled with air without tying off the end. The balloon will travel in the opposite direction of the motion of the air escaping the balloon.

Universal Law of Gravity

Isaac Newton puzzled over the same problem as Kepler in trying to understand what causes the planets to move in elliptical orbits around the Sun. The story goes that Newton, having watched apples fall from a tree, made the connection that the force of gravity, that acts on the Earth and causes objects to fall, is the same force that acts to keep planets in orbit around the Sun. At the time that the universal law of gravity was described by Newton in 1687, no one had suggested that gravity operates in space like it does on the Earth. Newton had unified the terrestrial with the heavens.

The universal law of gravity states that every object with a mass is attracted to every other object by the force of gravity. The strength of the gravitational attraction is directly proportional to the product of the masses, and inversely proportional to the square of the distance between them. For two bodies of mass M_1 and M_2 separated by a distance R, the force of gravity acting on either body by the other body is given by $F = G\dfrac{M_1 M_2}{R^2}$. Thus two bowling balls in deep space, separated by some distance R, will be attracted to each other. If we perform this experiment, we would find that the two balls would move toward each other until they collide. Since the strength of the gravitational force between two masses depends on the product of their mass, the gravitational force of attraction will increase (decrease) if the masses of either object increases (decreases). If the masses are held constant, but the distance between them is increased, then the force of gravity will decrease in strength. For example, if the distance between two objects doubles, the force of gravity will be one-fourth as strong. If the distance decreases by half, the force will be four times greater.

In our everyday experience, we see the force of gravity at work. It is the cause of objects falling, it is the reason we have tides, and it explains why the Earth, the Moon, and planets have the orbits that they do. If there was no gravity, life as we know it would not exist in our universe. It is a key part of what some have called "the fine-tuned universe."

Using both the universal law of gravity and the laws of motion, we can understand how a satellite can stay in orbit around the Earth, and by extension, help us to understand how a planet stays in orbit around the Sun. Naively, we would expect that since gravity is always an attractive force, the satellite and the Earth would be attracted to each other, resulting in the satellite falling toward the Earth. In reality, satellites can be made to orbit the Earth without falling directly toward its surface. How does this happen?

Figure 3.29 depicts a cannon placing a projectile into orbit around the Earth. Initially the cannonball is at rest inside of the bore of the cannon. When the gunpowder is ignited, the resulting expansion of hot gas imparts a force on the cannonball that causes it to accelerate (Newton's second law), and thus the cannonball moves forward. Since the cannonball is located near the surface of the Earth, there will be a strong gravitational attraction between the cannonball and the Earth, causing it to fall toward the Earth as it also moves in a horizontal direction due to the explosion of gunpowder. The net motion of the ball will be in the shape of a parabola that intercepts the Earth's surface, causing the cannonball to strike the Earth at position 1.

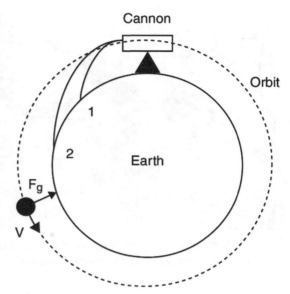

Figure 3.29 A satellite orbits Earth by acquiring sufficient speed to move sideways at the same rate it moves downward due to gravity. In essence, the satellite falls around the Earth.

If we repeat our cannon experiment with a larger amount of gunpowder, we find that the greater force on the cannonball from the explosion causes the ball to land further away from the cannon (position 2) than the first firing. If we could give the cannonball enough force from the ignition of the gunpowder, the cannonball would be projected forward with enough speed that it would fall toward the Earth at the same rate as the Earth curves away from the cannonball. This would cause the ball to make a complete circle of the Earth and thus be "in orbit." In practice, a cannon is not used to place an object in orbit, but a rocket is. We can design rockets so that they carry satellites above the frictional drag effect of the atmosphere and then propel them forward with enough speed so that they fall around the Earth.

As a satellite orbits the Earth, we know from Newton's second law that there must be a net force on the object since it is accelerating. Note that a satellite orbiting the Earth in a circle at constant speed is constantly changing direction. Since acceleration is a change in velocity (which has a speed plus direction component), when a satellite changes direction as it orbits it is also experiencing an acceleration. The force that causes the satellite to accelerate is the gravitational force between the Earth and the satellite. The gravitational force on the satellite due to the Earth is directed toward the center of the Earth. If the force of gravity were to mysteriously disappear, the satellite would fly away in a straight line at a constant speed (no acceleration), as Newton's first law describes.

It is instructive to note that if you were traveling inside of the satellite as it orbited the Earth, you would be weightless or in a "zero-g" environment. This does not mean that somehow gravity has disappeared. In fact, if there was no gravity, Newton's laws of motion state that the satellite would travel in a straight line at a constant speed in a path tangent to its orbit. The analogous situation arises

Source: NASA

Figure 3.30 Astronauts train for "zero gravity," but the term is misleading. Gravity is universal.

if you jump down an empty elevator shaft feet-first with a bathroom scale fixed to the bottom of your feet. Before you make the fateful leap, you notice that when you stand on the scale it indicates that you weigh 65 kg. As you fall down the elevator shaft you notice that the scale now reads 0 kg. Why are you weightless? You are obviously still within the gravitational field of the Earth.

What is happening is that the scale and you are falling at the same rate. Thus you are not pushing your feet against the scale as you fall down the shaft, hence no weight is measured. Applying this idea to the case of the satellite in Earth orbit, we see that both the satellite and you are falling at the same rate around the Earth. This gives the false impression that you are in a zero gravity environment. It is the reaction force (i.e., the "normal" force) caused by the gravitational force that is directed toward the center of the Earth that gives the feeling of weight or heaviness. For example, if you removed the floor that you are standing on you will not "feel" your weight as you fall downward. Thus something needs to push back on you in order for you to feel your weight (this is the normal force).

This line of thinking has deep physical meaning since if you equate Newton's second law with the universal law of gravity, we get $F = ma = G\dfrac{M_1 m}{R^2}$. Since the mass of the object that is falling (m) is on both sides of the equation, it will cancel to yield $a = G\dfrac{M_1}{R^2}$. Thus the acceleration that a falling object will experience is independent of its mass (only the Earth's mass, M_1, and the distance R from the object to the center of the Earth enter the equation). This explains why people and bathroom scales, as well as satellites in orbits, fall at the same rate.*

3.7 The Copernicans, Philosophy and the Nature of Science

In Chapter 1 you were introduced to scientist and philosopher, Dr. Charley Dewberry as you read his critique of the supposed objectivity and superiority of scientific knowledge. Here you will read more from his book *Saving Science*, as it pertains to the Copernicans and their role in the birth of modern science. While some historical revisionists attempt to convince their readers that the birth of modern science was predicated on abandoning religious faith, Dr. Dewberry will argue that the conflicts were not between science and religion, but between competing worldviews.

Were the Copernicans the Founders of Modern Science?

Figure 3.31 Kepler, Galileo, and Copernicus.

**The scientific community tells itself a marvelous story about the dawn of science, when the Copernican Revolution made it possible for mistaken and superstitious beliefs about the world to be replaced with real, objective knowledge about the world as it is. If, however, questions of science are inevitably subjective, does that mean I believe we should be skeptical about our ability to know the truth of things? Am I trying to say that the inevitable subjectivity of knowledge makes any claim to knowledge suspect and nullifies any supposed accomplishments of the Copernican Revolution? Not at all. Part of my problem with the currently popular view of science is that it fails to adequately consider the problem of skepticism and, in fact, ignores the true lessons of the Copernican Revolution. Unwittingly, the emphasis on "objective," peer-reviewed empirical science has undercut the ability of science to achieve its own goals and has violated its own ideals.

Two Questions

Both philosophers and scientists are interested in these two questions: Can we know the truth of things? And if can know, how do we know? Let us consider three possible answers to these questions.

1. Skepticism

One possible answer is skepticism: it is not possible to know what is true about the world. As I mentioned above, my own position might initially be suspected to be a skeptical one; if all knowledge is inherently subjective, why should we believe any of it? Why I am not a skeptic will become clearer later; I will only say now that I deny that recognizing the inevitable subjectivity involved in knowing leads us to skepticism.

The philosophies that have led to our modern understanding of science, however, cannot equally claim to avoid skepticism as their outcome. I have previously discussed how logical positivism, Popper's hypothesis testing, and Kuhn's views on paradigm shifts were all ultimately skeptical:

1. Logical positivists based their system of radical empiricism on the epistemology of Hume, which was inherently skeptical.

2. Popper did not renounce the skepticism of the logical positivists. His method involved an empirical test based on the criterion of falsifiability: we can never claim a theory is true; we can only claim it has been severely tested.

3. Kuhn, of course, by the very nature of his argument rejected the idea of the objectivity of scientific knowledge. He saw no objective criteria by which to pick one paradigm over another; we can never say one paradigm is closer to the truth than another.

Thus we see that the dominant philosophies of science in the twentieth century all lead to epistemological skepticism. This fact is very striking. Yet, whenever I talk to scientists about how they justify their beliefs, they always respond (if they respond at all) with some version or combination of these three philosophies, and most commonly they respond with the ideas of the logical positivists or of Popper. You would think, therefore, that these scientists would follow their philosophical roots to their natural conclusion and be skeptical about our ability to know true things about the world. But that is not the case. Instead, almost every scientist I talk to would give a second possible answer to our two questions.

2. Empirically Based Realism

My experience suggests that scientists are realists; they believe science tells them things about the real world. Yet, at the same time, they are strict empiricists. On the one hand, they believe in the superiority and objectivity of peer-reviewed empirical science—that is, they believe the three assumptions we have been discussing in this book, assumptions that grew out of the skeptical arguments of the logical positivists and Popper. But, on the other hand, they are not skeptics.

Frankly, I do not see how this can be. The only arguments I am aware of for the superiority of "objective," peer-reviewed science come straight from logical positivism and Popper. If those arguments are wrong (as I have argued) and if their skepticism is particularly unwarranted, then any justification for radical empiricism disappears. I can see no logically coherent way to be both a radical empiricist and a realist at the same time. Nonetheless, that is the position most scientists seem to have taken.

3. Empirically/Subjectively Based Realism

I would argue for a third answer to our two questions: a realism that recognizes the appropriate role of the subjective in knowledge. Empirical evidence is not bad; hypothesis testing is not bad; but in themselves they do not adequately account for how we know we can come to know the real world, but it requires more than the strict application of the scientific method; it requires reason, field experience, theory construction, historical evidence, and all those other subjective elements that science often rejects as "anecdotal" or "gray literature." I am concerned that by embracing the priority of peer-reviewed empirical science, the scientific community has at the same time unwittingly embraced implications that it does not believe.

The Copernican Debate

I will start at the inception of modern science. ***Examining the early development of science will help us understand the nature of science, its scope, and how its findings are justified.*** Because many of the assumptions of the early scientists differ from ours, examining them will help us both to illuminate our assumptions and to critique our position.

Because the Copernicans were instrumental in founding modern science, I will examine in some detail the debate that raged between the Copernicans and the Catholic Church concerning what science can know and what the grounds are for accepting that knowledge. In some sense, science has come full circle. I will argue that if today's empirical scientists and philosophers of science were to judge the facts of the Copernican debate, the overwhelming majority of them would side, surprisingly, with the Catholic Church against the Copernicans. This is particularly true of the empiricists and pragmatists who would most strongly reject the perspective for which this book argues. If I am right about the facts and their implications, the Copernican/Church controversy will help illuminate the central issues of my critique. In order for their view to be coherent the empiricists and pragmatists would need to rewrite history and deny the Copernicans their place as the founders of modern science.

My reason for returning to this historical controversy, however, is not to lobby for a revisionist history. What interests me is that all three of the dominant twentieth-century philosophies of science (the logical positivists', Popper's, and Kuhn's) agree that science can tell us nothing about the world. If their formulation of science and its methods were used to judge the Copernican debate, the Copernican position would lose to that of the Catholic Church. While it is beyond my task to defend the Copernican approach to science (although I think I could), I believe the Copernicans were largely right and the Catholic Church was wrong. For my purposes, however, if I can just show that the logic of the modern position is, in fact, similar to that of the Catholic Church and thus against that of the Copernicans, then I will have shown there to be serious problems with the modern position.

I will address two specific questions: Were the Copernicans (in particular, Copernicus, Galileo, and Kepler) warranted in claiming that their theory was true? And if they were warranted, on what basis did they warrant their claim? Hindsight on that historical battle shows us that the Copernicans' claims were true. At issue, though, is how the Copernicans defined science and how anyone can claim to know something in science.

To help clarify the debate, I will describe three likely positions that those involved in the debate held:

1. The Copernicans were not warranted in believing that their theory was true because the methods of astronomy and mathematics cannot prove what is true in the heavens. Their theory was only a method of calculating the appearances of the planets and the Sun, a method that might be useful but one that the Copernicans could not claim to be true.
2. The Copernicans were not warranted in claiming that their theory was true because they had not demonstrated its truth. It is possible for the methods of astronomy and mathematics to prove the true motion of the heavens, but the Copernicans did not successfully demonstrate their theory to be true.
3. The Copernicans were warranted in claiming that their theory was true because they had provided sufficient evidence to support their claim.

Those advocating position 1 obviously believed there was no warrant at all for the Copernicans claiming they knew the true motion of the heavens. Those advocating positions 2 and 3, however, agreed that proving the true motion of the heavens was possible, but they disagreed about whether the Copernicans had made their case. Why? Because they disagreed about what constituted a true warrant. As we will see in what follows, they emphasized (broadly speaking) either sense perception or reason as sufficient warrant for believing a theory to be true.

Now that I have described the three positions in the debate, let us look at the story of the Copernican Revolution as it is usually told today. I call it the "Copernican myth." Copernicus found astronomy on the brink of collapse, burdened by Ptolemy's complex geocentric system of epicycle upon epicycle. In a clean sweep, he replaced this geocentric nightmare with the simpler heliocentric system, thereby producing an accurate astronomy in far better agreement with reality. When Galileo took up Copernicus's cause, he ran into trouble with the obscurantist Catholic Church, which used the Scriptures and Aristotelian epistemology to try to defend its outmoded cosmology against the new scientific perspective based on sensory evidence and logical proofs.

Underlying this myth is the positivistic belief that science is the only form of knowledge. And so, according to the story, the Copernicans were warranted in claiming that their theory was true because, in keeping with the modern view of scientific knowledge, the Copernicans began with facts and then proceeded in a logical, inductive manner. In contrast, the Catholic Church grounded its opposition in religion, and there is insufficient warrant for religious belief (we are told) because faith is based on revelation and not ultimately on the evidence of the senses.

Upon examination, we will find that the Copernican myth is distorted in a number of crucial ways. None of the Copernicans were claiming that their heliocentric model made better predictions than those of the geocentric Ptolemaic system, nor were they claiming that they could prove their theory empirically. Nonetheless, they were claiming that their theory was true and that they had grounds for their claim. Furthermore, we will see that it is particularly ironic that so many twentieth-century scientists would champion the Copernicans, because none of the dominant twentieth-century philosophies of science (the logical positivists', Popper's, or Kuhn's) could agree that the Copernicans were justified in claiming their theory to be true.

So now, moving from the myth and keeping in mind the three positions on the debate that I described above, let us look at the Copernicans themselves and the views of the Catholic Church in order to answer our two questions: Were the Copernicans justified in claiming their theory was true? And on what basis did they warrant their claim?

© vkilikov/Shutterstock.com

Figure 3.32

Copernicus (1473–1543)

The first of the Copernicans, naturally enough, was Nikolaus Copernicus. Before we can talk about how he viewed the truth of and the warrant for his theory, we should clarify what he actually thought. Although it is true that he replaced Ptolemy's Earth-centered model of the solar system with a Sun-centered one, his thinking was somewhat different from what the Copernican myth portrays:

1. The system of epicycles in Copernicus's system was similar to Ptolemy's, and the system was quite rigid, not allowing for a number of additional epicycles.
2. Copernicus's system in no way increased the accuracy of predictions, Copernicus largely used Ptolemy's data.
3. The wanderings of the planets could be accurately accounted for by either system. Copernicus does not claim his system gives different empirical results.

We see therefore that Copernicus did not view his theory as simpler, better grounded empirically, or more accurate than the Ptolemaic system. This brings us then to our first important question: Did Copernicus believe his theory was true?

This is not an easy question to answer. The preface to his book, ***On the Revolutions of Heavenly Spheres***, clearly indicates that Copernicus did not believe his theory was true and, in fact, ascribes position 1 to him: because an astronomer cannot by any means calculate the true movement of the planets, Copernicus was only proposing a new system of calculations for computing the apparent location of the planets. But Copernicus did not write the preface. He only saw a printed copy of his work on the day he died. The preface appalled Galileo, Kepler, and several other followers of Copernicus. They clearly believed that Copernicus thought his theory was true—that he did indeed hold position 3. In his "Letter to the Grand Duchess Christina," Galileo claims that he should not like to have great men think that he endorsed Copernicus's position only as an astronomical hypothesis that is not really true. And at the beginning of his *New Astronomy* (1609), Kepler, likewise outraged, identified the Lutheran theologian Andreas Osiander as the person who inserted the preface. The preface, therefore, is not strong evidence for answering our question.

In *Revolutions*, Copernicus claimed that he was led to search for a new system for deducing the motions of the Sun, the Moon, and the planets for no other reason than the fact that astronomers did not agree among themselves. After years of study he proposed the Sun-centered solar system, and he concluded:

> Having thus assumed the motions which I ascribe to the earth later on in the volume, by long and intense study I finally found that if the motions of the other planets are correlated with the orbiting of the earth, and are computed for the revolution of each planet, not only do their phenomena follow therefrom but also the order and size of all the planets and spheres, and the heaven itself is so linked together that in no portion of it can anything be shifted without disturbing the remaining parts and the universe as a whole. (p. 508)

If Copernicus held position 1 as the preface indicated, then his position seems poorly reasoned. He would only have been claiming to have devised a new calculating method that yielded better results or was easier to use. Yet, nowhere does Copernicus make these claims, and so why should anyone prefer his calculating method to another's? Furthermore, the claim he does make about the coherence of his system does not seem relevant if Copernicus viewed his method as just a calculating device. The planets only *appearing* to move as a functional whole would have nothing to do with improved observations.

If Copernicus did not hold position 1, did he perhaps hold position 2? Since his theory was no simpler or more accurate at prediction than the one it replaced, why would he recommend that other astronomers accept it? He sets out to put astronomy on a solid footing. But if he thought he had not proven his case, why would he even publish his book? It does not seem likely that Copernicus held position 2.

We are left with position 3, which indeed seems to be the position Copernicus held. The text of *Revolutions* leaves little doubt that Copernicus believed his theory was true and not just a computing devise. First, nowhere does he claim his work is merely a hypothesis. Second, throughout his work he describes the actual motions of the planets. Third, in his dedication to Pope Paul III, Copernicus claims he will demonstrate that the Earth moves. Fourth, in the dedication, Copernicus ridicules the Church father Lactantius for speaking childishly about the Earth not being a globe; and yet, if Copernicus had just been proposing another (among many) computing hypothesis, he would not have run the risk of offending the Pope by criticizing a Church father. And fifth, Copernicus's close friends, Bishop Tiedemann Giese and Georg Joachim Rheticus, were appalled by the preface to *Revolutions*. There can be little doubt that Copernicus was claiming to describe the actual motions in the heavens.

Now we can turn to our second question. If Copernicus believed his theory was true, then on what basis did he believe himself warranted for thinking so? He did not base his claim on empirical observations of the individual planets, but rather he based it primarily on geometric reasoning (the arrangement of the orbit of the planets). He appealed to the coherence of the system, not to individual appearances of each planet. Copernicus's approach was an important shift in the methods of science. Although his system was empirically equivalent to Ptolemy's (nowhere did Copernicus claim his system yielded empirically different results or better predictions), Copernicus was claiming his theory explained the true motions of the heavens. He believed his theory was true, and the warrant for his claim was intellectual: his theory was *inherently rational*; "in no portion of it [could] anything be shifted without disturbing the remaining parts and the universe as a whole."

So then, we know Copernicus held position 3, but what position in the debate did the Catholic Church hold? The Church did not respond to Copernicus immediately. He was, after all, a Church canon who had helped in the monumental task of revising the calendar. In a book published not long after *Revolutions,* the Dominican friar Giovanni Tolosani wrote that Copernicus's theory was absurd because it was scientifically unfounded and unproven. But for over sixty years, besides Tolosani's critique, the Church paid little attention to Copernicus's theory and leveled few criticisms. We must look to Tolosani then to represent the Church's position. According to Tolosani, Copernicus did not have a physical theory from which he could deduce the Earth's motion and neither had he presented a cause for its motion. Furthermore, in keeping with Ptolemy and Thomas Aquinas who believed astronomy could only be probable knowledge, Tolosani thought Copernicus mistakenly believed astronomy and mathematics could provide a basis for cosmology and physics. Ptolemaic astronomy, based on Aristotle's philosophy, said we cannot know the true motions of the heavens; and the Thomists said astronomy and mathematics could not provide information about the inference from the effects to the cause. Therefore, if Tolosani's position represented the Catholic Church's position in the Copernican/Church debate, then we must say the Church did not believe Copernicus was warranted in believing his theory was true. At this point, then, we can say the Church rejected position 3, but whether the Church held position 1 or 2 is not clear.

Galileo (1564–1642)

Next we come to Galileo Galilei, who was the lightning rod in the controversy between the Catholic Church and the Copernicans. What position did he take, and what position did the Catholic Church defend? To determine their positions, we will focus primarily on two formal events, often referred to as the first and second trials. Surviving documents pertaining to both events can help us ascertain both Galileo's position and that of the Catholic Church.

The first trial occurred in 1616. The Holy Office had decided the Copernican theory was foolish and absurd philosophically and that it was partly heretical and partly erroneous theologically. The Church, therefore, issued two (conflicting) warnings to Galileo, but it did not prosecute him. The first warning forbade Galileo from holding the Copernican theory as true, but he could defend, teach, and discuss it as possible or even probable. The second warning forbade Galileo from holding, defending, or teaching the Copernican theory; but, it is important to note, he could still discuss it.

Source: NASA

Figure 3.33

The first warning, communicated to Galileo by Jesuit Cardinal Robert Bellarmine, was rooted in the epistemology of Thomas Aquinas, which makes a clear distinction between "true" knowledge and "possible or probable" knowledge. The first warning indicates, then, that the Church had placed Copernicus's theory into the category of possible or probable knowledge. In a famous letter to the Carmelite provincial Paolo Foscarini, Bellarmine said that if the Copernicans ever came up with sufficient proof that their theory was true, the Church would have to reinterpret Scripture; however, the Copernicans had not provided the necessary proof. This is position 2. Therefore, the Church allowed Galileo to defend, to teach, and to discuss the theory, but he could not hold it as true.

The second warning, coming from a faction within the Catholic Church, was rooted in a more skeptical epistemology. This faction, who strongly inclined toward mysticism, did not believe in the capacity of human intellect to attain truth. They held position 1: they did not believe the Copernicans could ever know if their theory was true. Therefore they not only recommended that Galileo not hold Copernicus's theory as true, but also that he not teach or defend it (although he could discuss it).

The second trial occurred in 1633, soon after Galileo had published *Dialogue Concerning the Two Chief World Systems*. The Church commenced proceedings against Galileo for violating the terms of the 1616 warnings. Again, within the Church there were two positions in the debate: the Dominicans held position 1, that the Copernican theory was to be understood as an abstract construct incapable of scientific proof; and the Jesuits held position 2, that the Copernican theory was possible knowledge but that it had not been proven.

The Church leveled two charges against Galileo: first, that he had intentionally concealed the more severe of the two warnings from his publisher; and second, that in his book he had blurred the distinction between "true" and "possible" knowledge. The first charge is not germane to my argument. The second charge is relevant, however, because in answering it, Galileo identified himself with position 2. He maintained that he had never held the Copernican theory as verifiably "true," but saw it as "probable" knowledge. He claimed that in an effort to be witty and clever, he had inadvertently made it appear that he accepted the Copernican theory as true. Galileo most likely believed the theory was true, but he apparently did not believe he had the necessary warrant to claim it was true. He penned this note in the preliminary leaves of his copy of the *Dialogue:*

> Take note, theologians, that in your desire to make matters of faith out of propositions relating to the fixity of sun and earth you run the risk of eventually having to condemn as heretics those who would declare the earth to stand still and the sun to change position—eventually, I say, at such time as it might be physically or logically proved that the earth moves and the sun stands still. (p. v)

Thus, while Galileo may have believed the Copernican theory was true, it appears he did not believe the Copernicans had yet proved it. Galileo clearly held position 2. But what warrant would Galileo have accepted for justifying the truth of a theory? Would it have been grounded in sense perception or in intellect?

According to the modern Copernican myth, a warrant acceptable to Galileo would have been grounded in sense perception, that is, in observation of the heavens. After all, wasn't Galileo defending the new empirical science against the dogmatic Catholic Church? And wasn't the new science based on experience and observation, in contrast to the Church's outmoded epistemology grounded in Aristotle and the authority of Scripture? While there is some truth in the conventional story, there are some problems with it, as well.

First of all, the Church did not deny observation was important. Its Aristotelian-Thomist epistemology began with sense perception and then followed with abstraction and ultimately to a discussion of causes. How did this differ substantially from Galileo's methods? Furthermore, although the Copernican myth claims the Churchmen refused to look through the telescope because they were afraid of the facts—and perhaps this was true of some individuals—this was certainly not true of all. Some individuals merely considered the telescope to be a faulty instrument; they viewed it as unreliable because when it was set to different lengths, the appearances changed. Cardinal Robert Bellarmine accepted the telescope and looked through it with Galileo, but Bellarmine came to different conclusions. And a number of individuals were likely astute enough to understand that little or nothing was to be gained by the telescope observations themselves; because while telescopes could provide some supporting evidence for the Copernican theory, they did not show parallax in the stars caused by the movement of the Earth. If the Earth moved, observation of the heavens should show some evidence of it, but no movement was observed. This lack of parallax was a major stumbling block for the Copernicans at the time, and the use of the telescope only compounded the problem. But not all Churchmen were afraid to look through it.

Second, it is not clear that Galileo himself began with observation. Certainly, Galileo incorporated into the theory his observations obtained by using the telescope, but these observations were not conclusive; Galileo did not rely on them alone. For instance, in an important exchange of letters between Galileo and Kepler, Galileo wrote to Kepler that he had been a follower of Copernicus for a number of years and had indubitable proofs that the theory was true, although he had hitherto not dared to defend the new system in public. When Kepler wrote back that he wanted to publish these proofs in German, however, Galileo broke off all correspondence with Kepler for twelve years. Furthermore, what observation did Galileo make that conclusively tipped the scales toward the Copernican theory? He did not point to a set of observations or to any one. So then, what role observation played in Galileo's science is not as clear as the Copernican myth sometimes assumes. In the *Dialogue*, the only positive argument Galileo tries to make for the Copernican theory is based on the tides (which was a wrong line of reasoning), and the rest of the *Dialogue* critiques the Aristotelian system.

It is hard to say what Galileo considered to be sufficient justification for the truth of a theory. It is not so clear what he meant by the "new way of observation and experiment." How do we reconcile the new method with the following famous quote from the *Dialogue*?

> Nor can I ever sufficiently admire the outstanding acumen of those who have taken hold of this [Pythagorean-Copernican] opinion and accepted it as true; they have through sheer force of intellect done such a violence to their own senses as to prefer what reason told them over that which sensible experience plainly showed to the contrary. For the arguments against the whirling of the earth which we have already examined are very plausible, as we have seen; and the fact that the Ptolemaics and Aristotelians and all their disciples took them to be conclusive is indeed a strong argument of their effectiveness. But the experience which overtly contradict the annual movement are indeed so much greater in their apparent force that, I repeat, there is no limit to my astonishment when I reflect that. Aristarchus and Copernicus were able to make reason so conquer sense that, in defiance of the latter, the former became the mistress of their belief. (p. 328)

Also in the *Dialogue,* time and time again, it is Simplicio (the Aristotelian) who defends the role of the senses in the acquisition of knowledge and who complains that in the Copernican system the senses must be denied, and it is Salviati (Galileo) who wishes to deny not only the principles of science, but sense experience and the very senses themselves.

Galileo is famous for the "experiments" that he "performed," such as rolling round objects down frictionless planes or dropping differently weighted objects in the absence of air friction. However, it is difficult to say exactly what role these experiments played in Galileo's view of what warranted belief. These experiments are hardly what come to mind when an empiricist talks about "the facts" obtained from experiments. For Galileo, observation was not merely the quantification of bodies, but something else. And his experiments seemed aimed not at convincing himself of the truth of his beliefs, but at convincing others.

Galileo also continued the development of a mathematical description of nature, but he did not use Kepler's discovery of the elliptical orbits. Galileo may not have understood the significance of Kepler's work. Or, as some evidence near the end of the *Dialogue* suggests (Galileo complains that Kepler got caught up in the Aristotelian concept of the occult; p. 462), Galileo may not have accepted Kepler's view of science.

So, our question remains: What warrant would Galileo have accepted for believing a theory to be true? At this point, we can say only two things: first, Galileo's view of a proper warrant is complex and difficult to determine; and second, his view was certainly not the usual view of today's empirical science.

Kepler (1571–1630)

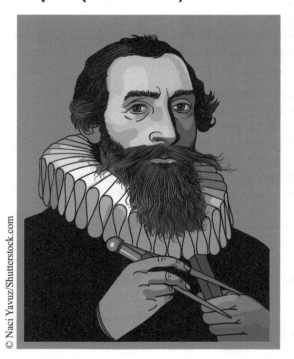

© Naci Yavuz/Shutterstock.com

Figure 3.34

Finally we come to our last Copernican example, Johannes Kepler, who published his *New Astronomy* in 1609. For a number of years, Kepler worked as an assistant to Tycho Brahe. Brahe had greatly improved the accuracy of the observations on the orbits of the planets, which Kepler used to discover these three laws:

1. The orbits of planets are elliptical in shape, with the Sun at one of the foci of the ellipse.
2. The areas of sweeps of a planet within their planes are always equal.
3. The ratio between the square of the periodic time and the cube of the distance from the Sun is the same for each planet.

There can be no doubt that Kepler held position 3: he believed the Copernican theory to be true. And there can be little doubt that Kepler's warrant for believing the theory true was based on the inherent rationality of the three laws and their correspondence with the observational data of Tycho Brahe. Kepler, like Copernicus, believed human reason could come to knowledge of the heavens. Kepler also agreed with Copernicus that geometrical/mathematical reasoning was an acceptable method of science. This is not to say, however, that observation was unimportant to Kepler. Without the observations Tycho Brahe made, Kepler could not have derived his three mathematical laws; but his warrant for accepting the Copernican theory was always based on the inherent rationality of the mathematical models. History seems to agree with Kepler about the relative importance of theory over empirical data: after all, it is Kepler whom we recognize as the great scientist and not Brahe, who collected the data.

Summary of the Positions

Now that we have looked at three prominent Copernicans, we can summarize the positions in the Copernican/Church debate. With these positions in mind, we can look specifically at how the Copernicans viewed science.

The Catholic Church

The Catholic Church's position in the debate was divided. The Dominicans held position 1: they believed that knowledge of the heavens was impossible. The Jesuits held position 2: they believed that it was possible to know the true motions of the heavens but that the Copernicans had not proved their theory.

The Copernicans

Copernicus and Kepler held position 3: they clearly believed that the theory was true and that they were warranted in claiming it was true because the system was inherently rational.

Galileo likely held position 3, but assigning him to only one position is difficult. He publicly claimed he held position 2. But there is significant evidence to show he believed that the Copernican theory was true and that the Copernicans were warranted in their claim; if he did, then he held position 3. We can say with some certainty, though, that Galileo did not believe strict empiricism to be the warrant in science.

The Copernican View of Science

Copernicus and Kepler saw science as a search for the true theories that explain the actual motions of the heavens. They were realists, motivated simply by the desire to know the truth. They believed that an individual could use his or her reason to come to know the truth and that their theory corresponded to the true motions of the heavens.

This view of science was the true watershed in the rise of modern science. Before the Copernicans, science looked to Ptolemy and Aristotle, who believed we could never know the true motions of the heavens. (They would have held position 1 in the Copernican/Church debate.) The Ptolemaic-Aristotelian view limits science to a small role in the acquisition of knowledge of the heavens.

Copernicus and Kepler believed they could use reason to come to know the true motions of the heavens, and the warrant for their belief was probably their view of God. They believed God was rational and that he had created the universe to be rationally ordered and human beings to be rational. And so the Copernicans believed they could use their rationality to know the truth about the created order—in particular, the true motions of the heavens.

Believing in man's rationality, the Copernicans did not limit science to strict empirical methods and justifications. Others at the time also believed that knowledge of the true motions of the heavens was possible, but they defined science as limited to more empirical or Thomist-type proofs.

Cardinal Robert Bellarmine, for example, took this position; he believed that the Copernican position was possible but that there was not sufficient proof for believing it to be true. In contrast, Copernicus believed his theory to be true based on the coherence between it and everything else he knew—in particular, mathematics. His theory was grounded in a geometric argument—not an empirical or Thomist argument based on the observable data. Both Copernicus and Kepler were relying on their rationality to arrive at a true theory, which Galileo recognized when he says (as I quoted earlier) that he cannot help but admire those individuals who hold the Copernican theory in spite of the data of their senses.

If science is limited to facts that can be empirically determined, then the Copernicans were wrong to claim their theory was true. This was the conclusion of Francis Bacon, the founder of modern empirical science. He argued that the Copernicans were wrong to claim their theory was true, because they had no empirical evidence.

The Modern View of Science

Attentive readers may have noticed that the three positions taken during the Copernican/Church debate correspond roughly with the three epistemological options (skepticism, empirically based realism, and empirically/subjectively based realism) discussed at the beginning of this chapter. In the light of our discussion of the Copernican Revolution, let us examine those three options again.

Skepticism

Ironically, the modern day skeptics have returned to the position that some in the Catholic Church held. Proponents of all three of the dominant theories of science (the logical positivists', Popper's, and Kuhn's) would have agreed with position 1 in the Copernican/Church debate: the Copernicans were wrong to claim their theory was true; based on their experience at the time, they could not know the true motions of the heavens. And so modern day skeptics would join the ranks of Ptolemy, Osiander (who wrote the preface to *Revolutions),* and some Dominicans in the Catholic Church—all skeptics who believed we could not know the true motions of the heavens. The positivists would have held position 1 because they are radical empiricists who explicitly accept the epistemology of David Hume and who therefore believe we can have no "objective" knowledge of the world outside of our minds. Those subscribing to Popper's view of science as hypothesis testing would have held position 1 because, believing that we can only falsify hypotheses, they deny we can know that a theory is true. And Thomas Kuhn would have held position 1; because of a problem he called "incommensurability," we cannot know if a particular theory (or paradigm) is closer to the truth than another theory. Anyone holding a conception of science that denies that science can come to the truth about the world would also have held position 1—along with one faction of the Catholic Church and against the Copernicans.

The only philosophical justification the scientific community has ever been able to provide for its radical empiricism leads inevitably to skepticism. The scientific community needs to confront the implications of this. If scientists can never know the truth, then on what basis should science have any status as knowledge, let alone an exalted status? None that I can see. And why should anyone pay for scientific research if it is not going to lead to true theories?

If any of the skeptical views are right, then science is dead. According to Kuhn, science only moves from one paradigm to the next and we cannot know if we have made any scientific progress; and even while a paradigm is in place, the scientist's work is only puzzle-solving. We cannot know if we are closer to the truth. And the positivists and Popper give us no surer foundation for science since they maintain that we cannot know what is true. Why should anyone bother to do science? Skepticism returns science to the cultural role it had in the Middle Ages.

Empirically Based Realism

Most scientists would agree with what I have said about skepticism; skepticism is not an option for practicing scientists. They may be empiricists, but they are not skeptics. Yet, as I argued above, the position of the realistic empirical scientist is not intellectually coherent. I do not see how one can be both a radical empiricist and a realist; the philosophical justification for each precludes the other.

For the sake of argument, however, let us suppose that somehow a scientist could embrace the empiricism of the logical positivists and Popper while rejecting their skepticism. Ironically, this would bring the scientist to the place where he or she would have agreed with those in the Catholic Church who held position 2—and so opposed the Copernicans. Anyone holding a conception of science that limits the warrant or justification of a scientific claim to strict empiricism would have denied the Copernicans' claim that their theory was true, because the Copernicans had not proven it empirically. Once sense perception is regarded as the sole basis for science, the Copernicans cannot be considered scientists. Again, this was Francis Bacons claim against the Copernicans.

Modern science only investigates questions that can be addressed by strictly empirical methods. But because these methods aim to replace intellect and reason with sense perception and statistics, using them results in limited, unskilled endeavors. These limited, empirical investigations are then published as peer-reviewed journal articles. And most scientists think that accumulating these small steps will help them decide between competing theories.

The method of science promoted by the Copernicans was substantially different from the modern conception of the scientific method. Because the Copernicans were evaluating the truthfulness of two empirically identical theories, no amount of direct empirical research could resolve the controversy. No amount of looking at sunrises could resolve the conflict. Data could not decide the issue. The Copernicans' method relied on reason and geometry—in short, on intellectual reasoning.

The modern scientific community needs to come to terms with the implication of strict empiricism. By truncating science to a supposedly objective, peer-reviewed, empirical process, modern science has nullified the accomplishment of the Copernicans. If radical empiricism is the essence of science, then modern science needs to rewrite its own history because those in the Catholic Church like Bellarmine were right and the Copernicans made a serious mistake. Now, I cannot believe the scientific community actually believes that the Copernicans were wrong. So I would like to see the scientific community revise its current view of science to include once again as scientists people like Copernicus, Kepler, and later theorists like Isaac Newton.

Empirically/Subjectively Based Realism

As must be apparent by now, I side with the Copernicans. Science is an enterprise where empirical data must take their place alongside the more subjective elements of reason and experience. Empirical data are important, but not uniquely so; Tycho Brahe's careful observations played a great role in the Copernican Revolution, but not a sufficient role. What modern empirical scientists seem to forget is that all data are linked to theories and to our previously accepted beliefs; we cannot investigate facts separated from their contexts. The development of theories is as important as the collection of facts. And in fact, we recognize most of the great scientists for their development of theories, not for their accumulation of data. If science ever forgets that the theorist is more important than the data collector, that Kepler's achievement was greater than Brahe's, it will truly have lost its way.

The Authority of Science

Many have concluded that the authority of science rests with the community of scientists through the peer-review process. The community ensures that the body of scientific knowledge contains only the knowledge it accepts as authoritative, and it does so through the peer-review process. I do not agree that the authority of science rests with the community of scientists.

To begin my critique, I will turn once again to history. During the transition from the Middle Ages to the Renaissance and the Enlightenment, the source of final authority shifted from the community of peers (the experts) to the individual. Luther's role in the Reformation and Galileo's role in the rise of modern science are two examples. The locus of science's authority became a central issue, especially in the Copernican/Catholic Church debate, as the Middle Ages declined and modern science rose. Who wore the final robes of authority for determining the truth of a theory? According to the Catholic Church, authority rested with the consensus of the community of natural philosophers. According to Galileo, however, every individual had the right to examine the evidence and decide for himself. Acknowledging this right was a significant, important, and good step; and because I do not believe many people would seriously disagree with this assumption, I will not defend it here.

I will return instead to the twenty-first century and ask this key question: Who now wears the final robes of authority concerning the truth of science? Those who insist on the peer-review process have placed these robes on the shoulders of a community of peers (the experts)—and have thus embraced a model of authority aligned more with the medieval Catholic Church than with the Copernicans. On the issue of authority, science has taken a giant step backward. The "community of science" has replaced the Church's "community of natural philosophers." One priesthood has replaced another.

By returning to an older model of authority, those holding the prevailing view of science have denied the progress made by the Copernicans, and Galileo in particular. Galileo's reply to one of his major opponents, Sarsi (a pseudonym for Jesuit Father Orazio Grassi), reveals his model of authority. When Sarsi argued against the Copernican theory, he implied that it must be wrong because it had attracted so few followers; even sixty years after Copernicus had published his work, the overwhelming majority of natural philosophers did not accept the Copernican theory. He cited

Jerome Cardan and Bernardino Telesio as examples of feeble philosophers who also had no follow-ers, deducing from their inability to attract supporters that their theories must be lacking. And so, Sarsi implied, if the Copernican argument were good, a large number of their peers would have accepted their theory, but since few had accepted it, the theory must be a poor one. Here is Galileo's reply to Sarsi's deduction:

> Perhaps Sarsi believes that all the host of good philosophers may be enclosed within four walls. I believe that they fly, and that they fly alone, like eagles, and not in flocks like starlings. It is true that because eagles are rare birds they are little seen and less heard, while birds that fly like starlings fill the sky with shrieks and cries, and wherever they settle befoul the earth beneath them. Yet if true philosophers are like eagles they are not [unique] like the phoenix. The crowd of fools who know nothing, Sarsi, is infinite. Those who know very little of philosophy are numerous. Few indeed are they who really know some part of it, and only One knows all.

> To put aside hints and speak plainly, and dealing with science as a method of demonstration and reasoning capable of human pursuit, I hold that the more this partakes of perfection, the smaller the number of propositions it will promise to teach, and fewer yet will it conclusively prove. Consequently, the more perfect it is the less attractive it will be, and the fewer its followers. On the other hand, magnificent titles and many grandiose promises attract the natural curiosity of men and hold them forever involved in fallacies and chimeras, without ever offering them one single sample of that sharpness of true proof by which the taste may be awakened to know how insipid is the ordinary fare of philosophy. Such things will keep an infinite number of men occupied, and that man will indeed be fortunate who, led by some inner light, can turn from dark and con-fused labyrinths in which he might have gone perpetually winding with the crowd and becoming ever more tangled.

> Hence, I consider it not very sound to judge a man's philosophical opinions by the number of his followers. Yet though I believe the number of disciples of the best philosophy may be quite small, I do not conclude conversely that those opin-ions and doctrines are necessarily perfect which have few followers, for I know well enough that some men hold opinions so erroneous as to be rejected by every-one else. But from which of those sources the two authors mentioned by Sarsi derive the scarcity of their followers I do not know, for I have not studied their works sufficiently to judge. *(Discoveries and Opinions of Galileo*, pp. 239–240)

Galileo is claiming that the individual has the final authority to judge the truthfulness of a work. His critic placed the final authority for accepting a theory within the community of workers (the experts); he believed that their ignoring the Copernican theory for sixty years was evidence of its being a poor theory. But Galileo rejected the notion that the number of workers supporting a theory indicated its truthfulness or worthiness.

Sarsi spoke for the Catholic Church's position, and his logic is clear: the experts were in a position to know. The issues were difficult, and individuals outside the community of experts were unable to critique this difficult philosophical work because they had not been trained to do so. Because Copernicus had been trained in mathematics and astronomy but not physics and logic, for instance, he did not understand that mathematics and astronomy could not arrive at true knowledge unless they were grounded in physics and logic.

The logic used to justify the peer-review process today is similar to Sarsi's. If peer reviewers accept a work, it must be a good study and it is accepted into the body of objective knowledge. If they do not accept the work, it must not be a good study and it is not accepted into the body of objective knowledge.

Therefore, the position of virtually everyone involved in salmon recovery has more in common with the Catholic Church's position than the Copernicans'. The prevailing view of science assigns the final authority for deciding whether a theory is true or not to the community of scientists, and the community executes this authority through the peer-review process: if a work is not peer reviewed, it is not objective knowledge. It seems ironic to me that in the twenty-first century, many would advocate a position for science that has more in common with the medieval Catholic Church than with the Copernicans'—especially when these same people would agree that moving the authority from the peer community (the experts) to the individual helped create modern science. They have simply replaced the Catholic clergy with a new scientific clergy. This is a giant step backward.

A Disclaimer

At this point, I need to clarify my position. I am not saying that peer-reviewed science is entirely fallacious and without any merit. After all, in the last one hundred years, science, largely through peer-reviewed journals, has achieved much. The problem I see rests largely with the expressed philosophy of science and its conclusions—not with the content of science. Scientists at their best are using a great deal of energy and skill to seek greater understanding of the world around them. But when they try to explain the philosophy behind what they are doing, and when they proclaim that scientific knowledge has greater standing than other forms of knowledge, I must object.**

Light and Telescopes

Chapter 4

> *"The wonderful relationships of the sun, the planets, and the comets could only have come into existence according to the plan and instructions of an omniscient and omnipotent Being."*

—Sir Isaac Newton
(inventor of the reflecting telescope)

*ASTRONOMY is unique among the sciences in the sense that the vast majority of objects that we study must be studied remotely. Except for meteorites, several hundred pounds of rock and soil from the Moon, and samples returned from a small number of objects such as Comet Wild 2 and asteroid 25143 Itokawa, astronomers predominately study distant objects via the electromagnetic radiation ("light") that they emit. This radiation travels through the emptiness of space over unfathomable distances. The analysis of this light can provide us with detailed information about stars and galaxies regarding their size, distance, speed, rotation, and chemical makeup.

Figure 4.1

All objects emit radiation in the form of electromagnetic waves.* That includes you! Right now you are giving off infrared radiation that we sometimes call body heat. *This radiation is a direct result of the vibration of charged particles within the emitting object. The average motion of particles in an object is characterized by its temperature. In general, the hotter an object, the higher its temperature, the higher the level of internal vibration or motion of particles, and the more light that it radiates. Although this statement is true in general, heat is not the same thing as temperature. The heat of a substance is a

---

*From *Learning Astronomy, 3/e* by Wayne A. Barkhouse and Timothy R. Young. Copyright © 2020 by Kendall Hunt Publishing Company. Reprinted by permission.
**From *Fundamentals of College Astronomy, 4/e* by Michael C. LoPresto and Steven R. Murrell. Copyright © 2019 by Michael C. LoPresto. Reprinted by permission.

Figure 4.2 Using an infrared thermometer.

measure of the total internal energy of the particles that make up the object. This includes the energy associated with the translational motion of particles (kinetic energy), rotational energy, and vibrational energy. Temperature is a measurement of the average kinetic energy of a system of particles.* Consider an ice sculpture and a lit match. Most people know that the lit match has a higher temperature, but it may seem odd that the ice sculpture actually possesses more "heat."

*To better understand the relationship between heat and temperature, recall that a pot of water at room temperature will increase its temperature as more heat is added. Once the temperature reaches 100°C, the water temperature will not increase, even though additional heat is added to the system. This additional heat is used to increase the rate of evaporation as the water (liquid) transforms into steam (gas).

4.1 The Speed of Light

In our everyday experience, light seems to travel instantaneously from one place to another. Over large distances the finite speed of light becomes noticeable. The speed of light in a vacuum (denoted by c) is assumed to be constant. Light travels through substances such as glass, water, and air, at a slower speed compared to the speed of light in the vacuum of outer space.

James Clerk Maxwell.

Figure 4.3

An interesting early attempt to measure c was done by Ole Roemer in 1676. Roemer measured the speed of light to be approximately 220,000 km/s by observing eclipses of the four brightest moons of Jupiter (Io, Ganymede, Europa, and Callisto). Observations showed that the eclipses of the Galilean satellites by Jupiter did not always occur when expected based on well-understood orbital mechanics. Roemer realized that the changing distance between Jupiter and the Earth would affect eclipse timings if the speed of light was finite. Based on these eclipse timings and an estimate of the distance between the Earth and Jupiter, it was simple to calculate the speed of light. In Roemer's time, the largest uncertainty in this technique was the accuracy of the estimation for the Earth–Jupiter distance.*

**In the late 1800s, physicist James Clerk Maxwell combined the basic equations that describe the nature of electricity and magnetism to theoretically predict the existence of *electromagnetic waves*. The speed of these waves, calculated from the basic constants of electricity and magnetism, is equal to about $c = 300,000,000$ m/s (3×10^8 m/s in scientific notation). That is about 186,000 miles per second! This is the speed of light. The symbol "c" is simply an abbreviation or shorthand that physicists and astronomers use for the speed of light.

Heinrich Hertz verified the existence of electromagnetic waves experimentally by producing them with vibrations of an electrified antenna. Because of his contribution, the unit for the frequency of waves is called the Hertz.** A.M. and F.M. radio stations are assigned their "channel" or call-number based on their frequencies. A.M. stations operate with kilohertz frequencies, and F.M. stations operate at megahertz frequencies.

Later, physicist Alfred Michelson measured the same speed for light waves as Maxwell's theoretical speed for electromagnetic waves. Michelson was the last in a line of scientists, beginning with Galileo, who had attempted to measure the speed of light, but Michelson's experiments were the most precise and the first that could be compared with and shown to match theoretical predictions. This meant that the speed of light waves was now known and their nature, that they were combinations of electric and magnetic fields, was now understood.

4.2 The Dual Nature of Light

Sir Isaac Newton proposed that light was made up of particles of energy that traveled in straight lines. Part of his justification for this claim came from a series of experiments that he performed with prisms. At the time his work seemed to disprove the view that light was a wave, but that conclusion would prove to be premature.

While Newton was experimenting with light and optics, a famous Dutch scientist, Christian Huygens, was also conducting experiments with light. He arrived at the opposite conclusion, namely that light had wave-like characteristics. Huygens published his views on the propagation of light in terms of spherical waves and wave fronts. This seemingly conflicted view of the nature of light was the beginning of what became known as the wave-particle duality of light. However, until the twentieth century there was no way to reconcile how light could be both a particle and a wave. For Newton and Huygens, it had to be one or the other.

*Generally speaking, a wave can be described by its wavelength and amplitude. Figure 4.5 depicts a transverse wave, a wave whose direction of motion is perpendicular to its displacement. The wavelength is described as the distance between adjacent crests, while the amplitude is the maximum displacement of the wave from its undisturbed position.

© Everett Collection/Shutterstock.com

Figure 4.4 Among his many accomplishments, Isaac Newton attempted to explain the nature of light.

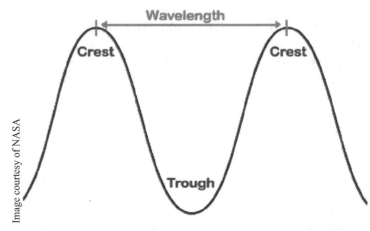

Image courtesy of NASA

Figure 4.5 A transverse wave depicting how wavelength is defined.

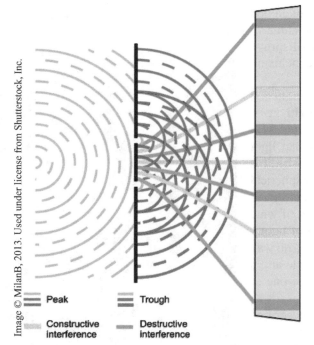

Image © MilanB, 2013. Used under license from Shutterstock, Inc.

Figure 4.6 Light passing through two narrow slits will combine in such a way to produce an interference pattern. A similar pattern can be seen when ocean waves pass through openings in a barrier.

In 1801, English scientist Thomas Young demonstrated conclusively that light had wave-like properties by noting that light, after passing through two narrow slits produced an interference pattern (Figure 4.6).* When wave peaks add together (combine) or when wave peaks and troughs cancel each other out, the effect is known as interference. *The interference pattern results from light waves constructively and destructively interfering as light from the two slits add together. The only way that such an interference pattern can form is that light must be made of waves. Many experiments conducted since 1801 have confirmed the wave-like nature of light.

In the late 1800s, Maxwell carried out a detailed study of electricity and magnetism, and formulated a set of equations that describe their properties. The equations that Maxwell formulated indicated that the effects of electricity and magnetism should travel together through space in the form of waves. The speed of propagation of these waves was determined to be equal to the speed of light. Maxwell suggested that these electromagnetic waves were in fact light waves, and this hypothesis was soon verified by experiment. We now know that light waves are waves of electromagnetic radiation that travel through space.

However, in 1905, Albert Einstein published a paper explaining why electrons are emitted from certain metals when light strikes their surface. This phenomenon, known as the photoelectric effect, was explained by suggesting that light consists of discrete bundles (particles) of energy (photons) that help liberate electrons via absorption of the light energy. Thus light having a simple wave-like nature could not explain the photoelectric effect.* By applying the particle nature of light to explain the photoelectric effect, Einstein earned a Nobel Prize.

Today we think of electromagnetic radiation as having both particle- and wave-like characteristics. Some experimental observations are best explained by thinking of light as a wave while others only make sense if light is a particle. It turns out that explaining the nature and behavior of light can be complicated, and our best models involve abstract mathematics. We attempt to use macroscopic terms that have a certain meaning to us ("wave" and "particle"), but neither provides a precise analogy for light.

4.3 Electric Charges

*To understand how electromagnetic radiation is generated, we need to review the structure of the atom. All matter is made up of different combinations of three fundamental particles: the proton, the neutron, and the electron. Both the proton and neutron are found in the nuclei of atoms, with the proton being positively charged and the neutron having no electric charge. The electron is found outside of the nucleus of an atom and carries a negative charge. Particles with opposite charges are attracted to each other via the electric force, while particles with like charges repel each other. The strength of the electric force depends both on the distance between charged particles and the amount of charge contained within each particle. The electric force between two point-like objects is given by Coulomb's Law: $F_e = \dfrac{KQ_1Q_2}{r^2}$, where Q_1 and Q_2 are the charges of particles 1 and 2, respectively, r is the distance between the charges, and K is the constant of proportionality. From Coulomb's Law, we see that the strength of the electric force between two particles is inversely proportional to the distance of separation squared, as is the case for the gravitational force of attraction between two objects.

Together, both electric and magnetic fields will be generated whenever a charged particle moves or vibrates. This combined electromagnetic field moves through space and is what we perceive as electromagnetic radiation or light (see Figure 4.7). It is important to note that it is the oscillating electric and magnetic fields that travel through space—not the charged particle—that are responsible for the electromagnetic radiation.

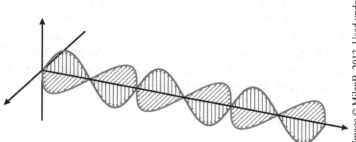

Figure 4.7 Oscillating electric (red) and magnetic (blue) fields are observed as electromagnetic radiation (light). Note that both the electric and magnetic fields are perpendicular to each other and the direction of propagation.

4.4 Properties of Light Waves

One of the basic properties of electromagnetic radiation is that all light waves travel at the same speed in a vacuum regardless of wavelength. In addition to characterizing light by its wavelength (λ-the Greek letter lambda), we can also describe light by its frequency (f), which is the number of complete cycles (waves) that pass by a given point per unit time. Normally, frequency is measured in cycles/second or hertz. Wavelength and frequency are inversely proportional to each other, and related to the propagation speed of the wave (v) given by $v = \lambda f$.

For light waves then, we can write this equation as $c = \lambda f$.

The energy per photon transferred by a wave is described by $E = hf$, where h is Planck's constant. Max Planck had proposed the idea of light's particle nature just a few years prior to Einstein's work on the photoelectric effect.*

Notice the resulting relationships based on these two equations. Radiation (light) with longer wavelengths must have lower frequencies and lower energy. Radiation with shorter wavelengths must have higher frequencies and higher energy. However, all forms of light travel at the same speed.

4.5 The Electromagnetic Spectrum

Visible light comes in a range of colors that are differentiated by their wavelength and frequency. The longest wavelength of visible light is red, while the shortest wavelength of visible light is violet. Wavelengths for visible light are usually measured in nanometers, where 1 nm is defined as 10^{-9} m. For the visible part of the electromagnetic spectrum, wavelengths range from 400 nm (violet) to 700 nm (red). You might be familiar with the colors of the visible spectrum as "ROYGBIV." Red has the longest wavelength, so it also has the lowest frequency and the lowest energy. Violet has the shortest wavelength, so it also has the highest frequency and highest energy for visible light. The other colors—orange, yellow, etc.—fall in between.

*The visible spectrum occupies only a small part of the total electromagnetic spectrum, whose wavelengths technically range from zero to infinity. Human eyes are only sensitive to wavelengths of light from 400 nm to 700 nm. Although the Sun produces radiation at other wavelengths, it is not equally bright across all wavelengths. In addition, the Earth's atmosphere acts like a filter that allows only certain wavelengths of light to pass through and reach the ground. The combination of the Sun's brightness as a function of wavelength and the effect of the Earth's atmosphere on absorbing electromagnetic radiation results in the Sun being the brightest from 400 nm to 700 nm as viewed from the surface of the Earth.

Since only a narrow slice of the complete electromagnetic spectrum is detectable by human eyes, we must build devices to allow us to observe wavelengths outside of the visible spectrum if we want to get a more complete picture when we observe celestial objects.

Image courtesy of NASA

Figure 4.8 The visible spectrum ranging from violet (400 nm) to red (700 nm).

The longest wavelength radiation is radio ($\lambda > 10$ cm), which includes radar and AM/FM and TV broadcasts. Microwaves are referred to as having wavelengths generally of the range 0.1 cm $< \lambda <$ 10 cm, while infrared radiation has wavelengths longer than red but shorter than microwaves. Although infrared light cannot be detected by our eyes, we are able to feel some infrared wavelengths as heat.

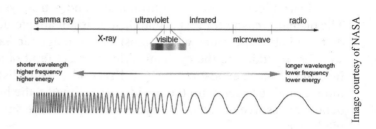

Figure 4.9 Electromagnetic spectrum.

Electromagnetic radiation with wavelengths shorter than visible light ($\lambda < 400$ nm) is known as ultraviolet light with 10 nm $< \lambda <$ 400 nm. Ultraviolet radiation is able to damage our eyes and skin. X-rays are generally designated as having wavelengths 0.01 nm $< \lambda <$ 10 nm. This radiation is highly penetrating and can cause severe damage to our cells. The shortest wavelength light, with $\lambda < 0.01$ nm, is known as gamma-ray radiation. Gamma-ray photons have the most energy per photon and can cause extreme damage to living tissue.

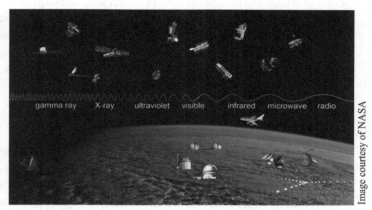

Figure 4.10 Astronomers use a wide variety of Earth-bound and space telescopes to make use of the full electromagnetic spectrum in their research.

4.6 Optical Telescopes

On a clear day or night, the Earth's atmosphere is mainly transparent to visible light. This allows us to see objects beyond the Earth's environment, such as the Sun, Moon, and stars. For wavelengths outside of the visible part of the electromagnetic spectrum, most of the light is scattered and absorbed by particles in the Earth's atmosphere. The degree to which light is scattered and absorbed is known as the opacity. The atmosphere is opaque at most wavelengths; thus shielding us from most of the electromagnetic radiation.* The two "windows" for Earth-bound observations are visible light and radio waves.

*The main contribution to opacity for visible light is clouds (optical telescopes cannot be used when it is cloudy). The ozone layer is responsible for absorbing ultraviolet, x-ray, and gamma-ray radiation. Both water vapor and carbon dioxide contribute to the absorption of infrared light.

Most objects studied by astronomers and astrophysicists are invisible to the unaided eye. The primary reason for this is that the eye is not able to detect light having a wavelength outside of the visible spectrum where most electromagnetic radiation is emitted from astronomical sources. In addition, the eye is unable to detect faint objects due to its small *aperture* or opening, and its short exposure time (~0.1 seconds). Thus distant galaxies, and even the planet Neptune, cannot be seen with the unaided eye. Finally, the human eye is limited in its ability to see details or resolve objects. The eye is only able to see surface features on nearby objects such as the Moon.

Major advances in astronomy over the past several hundred years have been driven by the development of instrumentation. The first device designed to improve our view of the sky was the telescope invented in the early 1600s. In 1609, Galileo Galilei became the first person to use a telescope to make astronomical observations and subsequently publish what he observed.

The primary goal of a telescope is to gather as much light as possible and bring it to a focus (a giant light bucket). With a telescope, the eye or some other detector, such as a camera, is able to observe/record the image of the object. There are two main types of telescopes that have historically been used in astronomy: 1) the refracting telescope, and 2) the reflecting telescope.

The Refracting Telescope

The refracting telescope, as the name suggests, is based on the principle of the refraction of light and was the type of telescope that was built and used by Galileo. When light passes through a medium it travels at a slower speed compared to the vacuum. For example, light travels slower through air than in a vacuum, and slower through water compared to air. As a result, light in general bends or refracts when traveling from one substance into another. The bending or refraction of light is how lenses are able to bring light to a focus, and illustrates how refracting telescopes, binoculars, microscopes, and even the human eye works.

The objective lens that is found in a typical refracting telescope is curved in such a way that light is refracted twice (once at the air/glass boundary and again at the glass/air boundary) so that it converges at the focal point. The lens is designed so that all light striking the surface of the lens will pass through the focal point. A detector, such as a digital camera, can be placed at the focal point to permit the recording of the image. Placing the eye at the focal point will result in an unfocused image since the eye has a lens too. To overcome this difficulty, an eyepiece consisting of a second lens is placed near the focal point so that light leaving the eyepiece will be parallel but concentrated. This allows the eye to bring light into focus on the retina; thus obtaining a focused image (Figure 4.11).

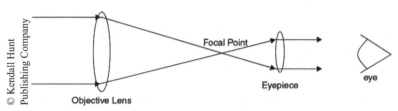

Figure 4.11 A drawing of a simple refracting telescope illustrating the direction that light travels to reach the eye.

The Reflecting Telescope

The reflecting telescope, first invented and used by Isaac Newton, uses a curved mirror to gather and bring light to a focus. This type of telescope is based on the principle of the reflection of light such that the angle of the incident light with respect to the normal (vertical line perpendicular to the surface of the mirror) is equal to the angle of the reflected light measured from the same normal. Reflecting telescope mirrors are coated with a reflective substance such as aluminum or silver. One interesting type of reflecting telescope consists of a rotating bowl of liquid mercury.

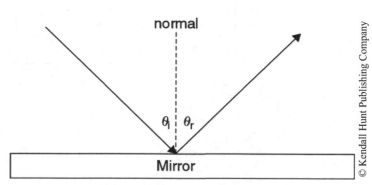

Figure 4.12 The law of the reflection of light states that the angle of incidence (as measured from the normal with respect to the mirror surface) is equal to the angle of reflection. This is the basic principle that describes how a reflecting telescope operates.

In order for light to be brought to a common focal point, the shape of the primary mirror must be curved to the shape of a parabola. If the surface of the mirror is in the shape of a partial circle, light rays striking near the edge of the mirror will not converge to the same location as light rays that hit near the central region of the mirror. Thus the image produced will be out of focus.

There are four basic types of reflecting telescopes that have been used in astronomy: 1) the Newtonian design, 2) prime focus, 3) the Cassegrain, and 4) Coudé focus. The Newtonian telescope was the first reflecting telescope to be developed and, as the name suggests, was invented by Isaac Newton. For this design, light is directed toward the side of the telescope tube after having been reflected from the primary mirror and a smaller flat secondary mirror located partially up the telescope tube (Figure 4.13). For the prime focus type, the detector is placed at the focal point of the primary mirror. This location is awkward since it requires placing a detector inside of the telescope tube. The Cassegrain type is the most popular type for professionals, and is designed such that light reflects off the primary mirror, back up the tube

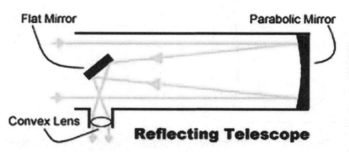

Figure 4.13 A Newtonian reflecting telescope consisting of a primary parabolic mirror and a flat secondary mirror. Note that the light gathered by the primary mirror is redirected to an eyepiece located at an opening on the side of the telescope tube.

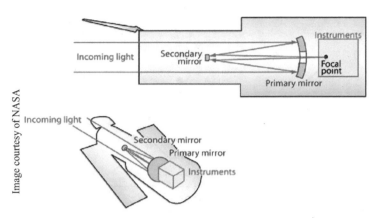

Image courtesy of NASA

Figure 4.14 The Cassegrain-type telescope is the general design that NASA used for the Hubble Space Telescope. Light enters the telescope, reflects off of the primary mirror, and then is redirected back toward the secondary mirror. Upon reflection from the secondary mirror, the light travels back toward the primary mirror and through an opening in its center.

to a convex secondary mirror, and back down toward the primary mirror. The reflected light then passes through an opening in the primary mirror and then into a detector (Figure 4.14). The advantage of this design is that the heavy instrumentation is located directly behind the primary mirror, thus making it relatively easy to balance the telescope as it tracks objects across the sky. The Coudé focus telescope is set up with two secondary mirrors that redirect light reflected from the primary mirror to the side of the telescope tube. This design is similar to the Newtonian reflector, and is used for very heavy instrumentation such as large spectrographs.*

Refractors vs. Reflecting Telescopes

Near the end of the nineteenth century, the last of the large, research refracting telescopes was built for the Yerkes Observatory in Wisconsin, with a record 40-in diameter lens. *All modern day telescopes are reflectors rather than refractors for several reasons. First of all, one disadvantage of a refracting compared to a reflecting telescope is that light is required to pass through the glass objective lens. As light passes through the lens, some of the photons get scattered and absorbed, resulting in a reduction of the amount of light that reaches the detector.

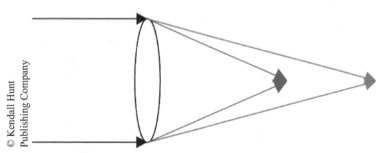

© Kendall Hunt Publishing Company

Figure 4.15 The drawing illustrates the effect of chromatic aberration that affects refracting telescopes. Light with a short wavelength will be refracted more than light with a longer wavelength as it passes through the objective lens. The result is that the location of the focal point (focal length) will depend on the wavelength of light. Thus blurry images will result if the light from an object contains multiple wavelengths.

Also, refracting telescopes suffer from an optical condition known as chromatic aberration. This effect is due to the fact that the refraction of light is wavelength-dependent. For example, blue light will be refracted more than red light, resulting in blue light coming to a focus at a shorter distance from the objective lens than red light. Thus, when observing an object that emits light over a range of wavelengths, the object will not be in focus for all colors. One method to correct this aberration is to include a second lens behind the first such that light passing through the second glass element will allow all light to come to a common focus point. The disadvantage of this correction is the cost of manufacturing a corrective lens, and the additional absorption of light due to the addition of a second lens in the optical path of the incoming light (recall that the primary goal of a telescope is to gather as much light as possible and bring it to a focus).

An additional problem with refracting telescopes is the difficulty in supporting the weight of glass objective lenses. Since light needs to pass through the front and back surfaces of the lens, the lens must be supported on its edge. Due to the very large weight of a typical glass objective lens, the shape of the lens will distort and sag slightly. This causes the light passing through the lens to diverge from a common focal point, thus producing a blurry image.*

Finally, as the size of lenses increase, refracting telescopes become much more costly, require much longer tubes, and create greater issues with the total weight of the telescope. By comparison, reflecting telescopes can be built less expensively, and with more compact bodies. For the amateur astronomer who might want to transport a telescope to different locations, these are important considerations. For professional astronomers, attempting to build larger refracting telescopes became completely impractical.

Telescope Light Gathering Ability

*One of the most important functions of a telescope is the ability to collect light. This light gathering ability is dependent upon the area of the primary mirror or lens of the telescope. The size of a telescope is usually given as the diameter (D) of the primary mirror or lens. The area (A) is proportional to the square of the diameter and is given by $A = \frac{\pi D^2}{4}$.

Thus a 3.0-m diameter telescope gathers nine times as much light as a 1.0-m diameter telescope.

The typical size of backyard telescopes is on the order of 0.2 m or so. The smallest professional telescopes in use are typically 1 to 2 m in diameter. During the 1980s, 4-m class telescopes were the largest in use. This includes the 4.0-m Mayall telescope at the Kitt Peak National

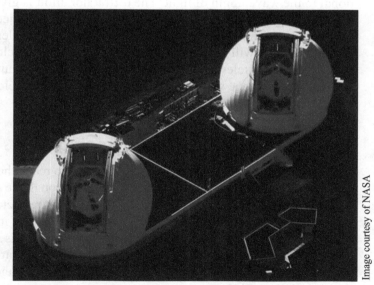

Image courtesy of NASA

Figure 4.16 The twin 10-m Keck telescopes are located on top of Mauna Kea, Hawaii at an elevation of approximately 4,100 m. Both telescopes have a primary mirror made up of 36 hexagonal single mirrors that are together in a mosaic to act as a single mirror.

Observatory in Tucson, Arizona, and the 4.0-m Blanco telescope in Chile. Today, the largest telescopes are 8 m in size, with some telescopes in the 10-m class such as the twin Keck telescopes on Mauna Kea, Hawaii and the 10.4-m GTA telescope on the Canary Islands.

Plans are in the works today to build even larger optical telescopes on the order of 30 to 40 m in diameter. One of the major motivating factors in wanting to build ever larger telescopes is the desire to see fainter objects. Increasing the light gathering ability is a way to accomplish this goal.* The use of multiple segmented mirrors acting together as a single mirror has allowed engineers to build such large telescopes.

4.7 The Resolving Ability of a Telescope

*Besides improved light gathering ability, a larger telescope is also better able to resolve objects that are close together and thus see finer detail. Every telescope has a fundamental limit on its ability to separate closely spaced objects. This limit, known as the diffraction limit, arises from the wave-like characteristics of light, and thus affects all optical telescopes. Light passing through a barrier (i.e., the aperture opening of a telescope) will tend to spread out in a diffraction-like pattern. A telescope with a small diffraction limit will be able to resolve finer details, while a telescope with a larger diffraction limit will produce images which look blurry compared to those produced from a telescope with a smaller diffraction limit.

A telescope's diffraction limit is inversely proportional to its diameter and directly proportional to the wavelength of light. The diffraction limit is given by $\theta \sim \lambda/D$, where θ is the angle of separation of two closely spaced but barely resolved objects, D is the diameter of the primary mirror or lens, and λ is the wavelength of light. Larger telescopes will have a smaller diffraction limit and hence will be able to resolve objects that are closer together than those observed by smaller diameter telescopes. Thus to maximize the amount of detail that can be observed for a particular object, one would choose the largest diameter telescope possible operating at the shortest wavelength.

The Effect of Atmospheric Seeing

In practice, ground-based telescopes are not able to reach their diffraction limit due to the blurring effect of the Earth's atmosphere. Light traveling through the atmosphere passes through air pockets of slightly different densities, resulting in light being refracted at slightly different angles. The end result is that light takes a slightly different path length as it travels through the atmosphere. This results in a blurring of the object image and is referred to as *seeing*. Seeing is quantified as the size of the image that a star will produce. The smaller the stellar image, the "better" the seeing. When air turbulence is high, the atmosphere will degrade the image quality to a greater extent than when the air turbulence is less.

In general, seeing varies from location to location and also as a function of time. One factor that affects seeing is the amount of atmosphere that the light passes through. When looking close to the horizon, seeing is generally worse than when looking toward the zenith position. Also, seeing is generally better when observations are taken at higher altitudes compared to sea level observations. Seeing is also worse when there are temperature differences between the air surrounding the telescope inside the observatory dome and the ambient air outside of the enclosure.

Several methods are used to help reduce seeing for ground-based observations and thus minimize the blurring effect of the atmosphere. These methods include placing telescopes in specific regions where the atmosphere tends to be more stable and having a larger fraction of nights with good seeing. Also, telescopes are placed at high altitudes to get above a significant portion of the atmosphere in order to help reduce air turbulence above the telescope. Observatory domes are designed and operated so that the temperature difference between the outside and inside air of the telescope enclosure is minimized. Seeing can also be kept to a minimum by looking at regions close to the zenith position rather than near the horizon. All of these methods are used to help reduce the effect of the atmosphere on the quality of images that are obtained by ground-based telescopes.

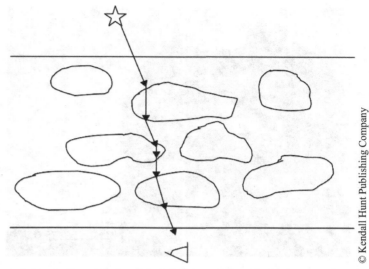

Figure 4.17 The light from an astronomical source changes direction multiple times as the light ray passes through various pockets of slightly different densities. The net effect is a blurring of the source image.

© Kendall Hunt Publishing Company

Improving Ground-Based Resolution

To compensate for the blurring effect of the Earth's atmosphere, several techniques have been developed to improve the resolution of ground-based telescopes. The oldest of these methods is known as speckle interferometry. Speckle interferometry is based on taking a series of short exposures so that the atmospheric seeing changes very little during a single exposure. Once a sufficient number of exposures have been obtained, the final image is formed by stacking together the individual images. Any shift in the location of the image due to seeing effects can be corrected for by shifting individual exposures so that the image lines up for each exposure when making the final image. The drawback of this technique is that individual exposures must be kept short to ensure that the atmosphere changes very little during each individual exposure. In the past, only bright objects could be detected in a single exposure, and hence speckle interferometry was not used to image faint sources.* However, recent improvements have led to better imaging. Think of it as speckle 2.0 for ground-based telescopes.

Image courtesy of NASA

Figure 4.18 The Gemini South telescope in Chile has seen big improvements with imaging thanks to new speckle instruments.

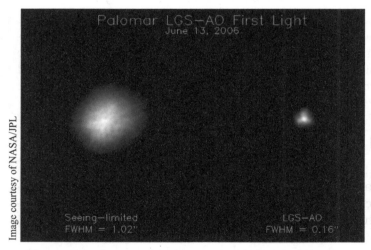

Image courtesy of NASA/JPL

Figure 4.19 Image of a star taken with (right) and without (left) the Palomar adaptive optics system. The image resolution is improved by over a factor of 6 with active compensation of atmospheric seeing.

*With the development of small, powerful, and affordable computers in the 1980s, astronomers and engineers designed computerized systems that actively tilted a telescope's primary mirror many times per second in order to compensate for the slight change in direction of wave fronts passing through the atmosphere from an astronomical source. This system of active optics (or "adaptive" optics) adjusts for the change in direction of the incoming light by monitoring light emitted from atmospheric sodium atoms that are excited by a ground-based laser. This technique works by knowing the light profile expected from the sodium atoms and making adjustments to the tilt of the mirror so as to reproduce the expected image.

As active optics systems were developed, the method expanded to incorporate the active deformation of the primary mirror using a system of hydraulic pistons that push and pull the primary mirror ever so slightly at different points on the non-reflective side. By actively controlling the shape of the mirror by performing many adjustments per second, light waves striking any part of the mirror would be focused more precisely than without using the adaptive optics system.

Space Telescopes

Although great progress has been made in reducing the seeing effect of the Earth's atmosphere on ground-based observations, placing a telescope in space above the atmosphere is the best option. There have been many telescopes placed in space in either Earth orbit or in orbit around the Sun. The ability of these space-based telescopes to resolve closely separated objects is limited only by the optical system of the telescope (i.e., the diffraction limit).

Besides resolution, the other advantage of placing a telescope in space is to avoid the complete absorption of certain wavelengths of light by the atmosphere. Observing astronomical sources from space allows us to sample the complete wavelength range of the electromagnetic spectrum.

A disadvantage of placing a telescope into space is the cost. Not only are funds required to build and operate a telescope, there is also a significant cost associated with launching a telescope into space.

The most famous space telescope ever built is the *Hubble Space Telescope*, which was launched from the space shuttle *Atlantis* in 1990.

Image courtesy of NASA

Figure 4.20 Image taken of the Hubble Space Telescope during the servicing mission 3B in March 2002.

Named after Edwin P. Hubble, an American astronomer , the telescope is a 2.4-m diameter Cassegrain-type reflector and observes from ultraviolet to infrared wavelengths.* When it was launched, *Hubble* had an expected lifespan of about ten to fifteen years. However, due to some initial problems with the telescope, it became clear that astronauts in the shuttle program could make repairs and updates. Thanks to work done in five different visits to the school bus-sized telescope, *Hubble* has had a long, extended, successful career. *The *Hubble Space Telescope* has revolutionized optical astronomy with its high-resolution imaging of planets, stars, and galaxies.

Astronomical Detectors

After light travels through a telescope, it can be collected and stored using a variety of methods. The use of an eyepiece allows one to see visible light images created by a telescope. This is a popular method for backyard astronomical viewing, but does not allow for the recording of observed images. Since the eye accumulates photons for only a brief interval of time (~0.1 seconds), many faint objects are invisible to the unaided eye. In addition, the eye is only capable of detecting visible light; thus most of the electromagnetic spectrum would go unobserved if we only use our eyes to view.

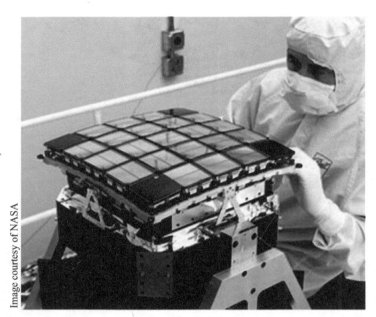

Image courtesy of NASA

Figure 4.21 The focal plane of the Kepler space telescope showing all 42 CCDs that make up the primary imager.

A major breakthrough in recording astronomical objects came during the 1800s with the invention of photography. Photographic plates or film allows for the recording and storage of observations in an essentially permanent fashion. The photographic plate is placed at the focal plane of the telescope (the focal length distance from the primary mirror or lens), with the resolution limited by the size of the grains that make up the photographic emulsion. The major advantage of the use of photography for astronomy is the ability to control the exposure time. Taking long exposure images allows for the recording of very faint objects that are invisible to the human eye. Thus vast regions of the universe were unveiled with the invention of photography.

In the 1980s, the use of charged couple devices (CCDs) became commonplace at all major observatories. Today CCDs have replaced traditional photographic film and plates not only in astronomy but also in the consumer marketplace in the form of digital cameras. CCDs have many distinct advantages over photographic film, the major one being that CCDs are more sensitive to faint light than film or plates. For example, a traditional photographic plate records approximately 1 to 2 photons for every 100 photons that strike the emulsion, while a CCD typically records 99 of 100 photons that strike the silicon chip. In practice this means that for an equivalent exposure time, a CCD is able to record fainter objects than photographic film or plates.

CCDs are also capable of achieving higher resolution than photographic plates or films. This can be done by manufacturing CCDs to have small pixels (or picture element), which are single light sensitive elements in the array that comprise a CCD. The resolution element for a photographic plate or film is limited by the size of the emulsion grain. CCDs can be constructed to have pixels that are several times smaller than the resolution element of a photographic plate. Thus CCDs can record finer details of an object compared to using photographic plates or films.

An additional advantage of a CCD is that they are sensitive over a broader wavelength range than photographic plates or films. Also, since the output from a CCD is digital in nature, data can be easily stored, transported, and analyzed by computer software. All modern astronomical observations are conducted by CCDs.

4.8 Astronomy Across the Full Spectrum

Radio Astronomy

If we restrict ourselves to observing the universe around us to visible light, we will obtain a very biased understanding of the cosmos, since most of the electromagnetic radiation emitted from astronomical sources occurs at wavelengths outside of the visible spectrum. Observing non-visible light requires us to invent telescopes and detectors that are sensitive to various wavelength regions that cannot be observed with optical telescopes.*

Decades after Marconi demonstrated that radio waves could be sent and received as a new means of communication, we learned that there are radio waves from objects in space that reach the surface of the Earth. *These radio waves were first detected in 1931 by Karl Jansky using an antenna designed to rotate in any direction. Jansky measured a "noise" that tracked across the sky in the same way expected for stars (i.e., sidereal time). After careful study, Jansky realized that the source of the radio waves that he was detecting was located near the center of the Milky Way galaxy in the direction of the constellation Sagittarius. This was the beginning of the discipline of radio astronomy.

Since radio waves have much longer wavelength ($\lambda > 10$ cm) compared to visible light (400 nm $< \lambda <$ 700 nm), there are many design differences between radio and optical telescopes. Recalling the expression for the diffraction limit of a telescope, $\theta \sim \lambda/D$, radio

Figure 4.22 Astronomers study objects like the Sun across a full spectrum of electromagnetic radiation to get a more complete picture.

Figure 4.23 Multiple views of the Milky Way galaxy, taken at a variety of wavelengths.

Image courtesy of NASA

telescopes need to have a very large diameter in order to obtain a reasonable resolution. In practice, single dish radio telescopes have a poorer resolution than their optical counterparts.

An advantage of radio telescopes compared to optical telescopes is that radio telescopes do not need to be as smooth as an optical mirror or lens. Since the wavelengths of optical light are on the order of a few hundred nanometers (billionths of a meter) in size, optical mirrors and lenses need to be smoothed to near perfection. Since radio waves are larger than a few centimeters in size, the surface of a radio telescope does not need to be as smooth as an optical telescope. This saves money during the construction phase.

Due to the very large diameters of radio telescopes compared to optical telescopes, radio telescopes are very sensitive to faint radiation. This allows radio telescopes to detect astronomical sources that may not be visible to optical telescopes, even though the source may be of equal brightness at radio and optical wavelengths.

Radio astronomy has an advantage over optical astronomy in that radio observations are able to be conducted during the day and night, even under cloudy conditions. At radio wavelengths, the sky is dark and thus the atmosphere does not affect the transmission of radio waves compared to visible light. The reason for this is that gas and dust particles in our atmosphere are of the right size on average that short wavelength blue light from the Sun gets scattered preferentially compared to the longer red light. As the light from the Sun passes through our atmosphere during the day, a significant fraction of the blue light is scattered about while the longer wavelength light takes a more direct path to our eyes. The net effect of preferentially scattering short wavelength visible light is that the eye sees blue light coming from every direction in the sky during a clear day. This light is bright enough to overwhelm the light from the stars and galaxies, thus making them invisible to the unaided eye, and giving rise to our blue sky. In contrast, radio waves pass through the Earth's atmosphere without any appreciable scattering or absorption (we are ignoring the fact that some very long wavelength radio waves are not able to pass through the Earth's atmosphere). Thus at radio wavelengths, the sky is dark during the day, and it makes no difference if one observes during the day or night. What can affect radio observations is the presence of a nearby thunderstorm. The electrical discharge from the lightning can cause interference with detected radio signals, thus most radio observations are suspended during such conditions.

One major advantage that long wavelength radio emission provides is the ability to detect electromagnetic radiation from sources that are

Image courtesy of NASA

Figure 4.24 The 100-m Green Bank Telescope.

embedded within regions containing a high density of gas and dust. Optical light is absorbed and scattered by interstellar material, thus making it impossible to observe what is happening inside of these clouds if the density is high. Radio waves are able to make their way out of these dense regions, thus permitting us to observe the physical processes at work in these regions.

The 100-m Robert C. Byrd Green Bank Telescope is the world's largest fully steerable radio telescope. The telescope is located in Green Bank, West Virginia and is part of the National Radio Astronomy Observatory.* Like all radio telescopes, the Green Bank Telescope must deal with radio interference from outside sources. Just as optical telescopes may be plagued by light pollution, radio telescopes must deal with ground-based interference. Computers, microwave ovens, garage door openers—any type of electronic device—and even sparkplugs firing in a vehicle's engine will produce stray radio waves that can ruin the data astronomers are trying to collect.

The largest non-steerable radio telescope had been built in Puerto Rico at Arecibo. The Arecibo telescope was over 300 m across, and served astronomers well for decades. However, it fell into disrepair and then collapsed in 2020. Today the largest radio telescope is in China, with a monstrous 500-m dish.

Radio astronomy has given us a number of important discoveries, including the existence of pulsars and quasars, along with more "local" information about the Sun and other solar system objects.

Radio Interferometry

*As emphasized previously, one of the main hindrances of radio compared to optical observations is the poor resolution obtained by radio telescopes due to the large wavelength of radio waves. Scientists have developed a very powerful technique called interferometry that allows astronomers to overcome the problem of poor resolution when making radio observations. Interferometry works by having several independent radio telescopes observe the same astronomical object. The spacing between the individual telescopes serves as the effective "diameter" of the telescope. For example, the Very Large Array (VLA) in Socorro, New Mexico consists of 27 radio telescopes spread over an area of 30 km across. The VLA acts as one giant radio telescope having a diameter of 30 km. The diffraction limit in this case is reduced by a factor of 1,200 when D is 30 km rather than 25 m for an individual VLA radio telescope.

The technique of interferometry has been extended to using several radio telescopes spread over the surface of the Earth.* For example, astronomers have paired up radio telescopes in Hawaii and the Virgin Islands. *In this case, the effective diameter of the radio telescope is the size of the Earth. The huge reduction in the diffraction limit

Figure 4.25 The VLA in New Mexico.

at radio wavelengths using interferometry has allowed radio astronomers to achieve resolutions that are often superior to ground-based optical observations.* With our Earth-sized example, imagine a telephoto lens that could take a picture of a baseball that is sitting on the Moon!

*One of the most sensitive interferometers ever built, the Atacama Large Millimeter Array (ALMA), is in Chile. ALMA consists of 66 radio telescopes, each 12 m in diameter.

Radio observations can map out the all-sky distribution of atomic hydrogen in the Milky Way. The 21-cm radiation from hydrogen clearly outlines the disk of our galaxy in a bright band. Radio observations of Sagittarius A, located near the center of the Milky Way, show a large filamentary-like structure.*

Infrared Astronomy

In 1800, astronomer William Herschel did experiments that led him to the discovery of infrared light. We can't see this light, but sometimes we can sense it as heat. Night-vision goggles are used to detect this heat signature, and if you ever watched the old "Predator" movie you know that is how the alien hunted his prey. Your television remote likely uses infrared waves as well. Herschel found this radiation just below red on the visible spectrum, thus the name infrared.

*The study of astronomical sources using infrared light (700 nm $< \lambda <$ 10,000 nm) allows astronomers to uncover physical processes at work for relatively cool objects, those that are enshrouded by interstellar material, and objects that are very distant. Earth-based infrared study is extremely limited. The Earth's atmosphere is not transparent to all infrared wavelengths since a lot of the radiation is blocked by water vapor and carbon dioxide. Although infrared observations can be made from the ground at some wavelengths, infrared telescopes are usually placed at high altitudes or in space to minimize the absorbing effect of the atmosphere. Space-based telescopes are the preferred instruments to use for observing the universe at infrared wavelengths.

The first space-based infrared telescope that revolutionized astronomy was placed in Earth's orbit in 1983. The *Infrared Astronomical Satellite* (*IRAS*) was the first to conduct an all-sky survey at infrared wavelengths, and it made many discoveries related to star formation, interstellar matter, comets, and galaxies.

In 2003, NASA launched the *Spitzer Space Telescope* as the follow-up mission to *IRAS*. The superior detectors and larger primary mirror of *Spitzer* allowed for higher resolution

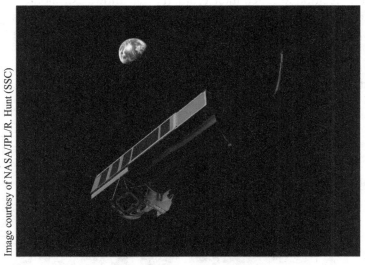

Image courtesy of NASA/JPL/R. Hunt (SSC)

Figure 4.26 Artist conception of the Spitzer Space Telescope ejecting its dust cover as it started its mission. Spitzer is designed to observe the universe from 3 to 180 microns (1 micron = 1,000 nm).

and greater sensitivity than could be achieved by *IRAS*. *Spitzer* has helped us make tremendous progress in understanding various astronomical sources and physical processes at work in our universe.

In 2021, the replacement for the *Hubble Space Telescope* is expected to be launched. (It has been delayed multiple times.) The *James Webb Space Telescope* (*JWST*) will have a primary mirror 6.5 m in diameter, nearly three times larger than the *Hubble Space Telescope*. In terms of light gathering ability, *JWST* will gather more than seven times as much light as the *Hubble Space Telescope*. The detectors on the *JWST* are designed to be most sensitive to infrared light, while the spacecraft has a large sunshield to help prevent overheating of the instruments from the Sun which could swamp the signal coming from astronomical objects. The telescope will be placed in a special orbit around the Sun approximately 1.5 million km from Earth, known as the L2 or Lagrangian point. At this location, the combined gravitational effect of the Sun and Earth on the spacecraft will cause the telescope to orbit the Sun with the same period as the Earth. This orbital position is highly stable and thus requires minimal energy to maintain the spacecraft at its target location. The *JWST* is expected to revolutionize astronomy and astrophysics, as it will be the largest infrared telescope ever placed in space.*

Image courtesy of NASA/JPL-Caltech/K. Gordon (University of Arizona)

Figure 4.27 A Spitzer image of our nearest galactic neighbor, the Andromeda Galaxy (M31). Due to the relative closeness, we are able to trace the spiral arm structure right to the central region of the galaxy.

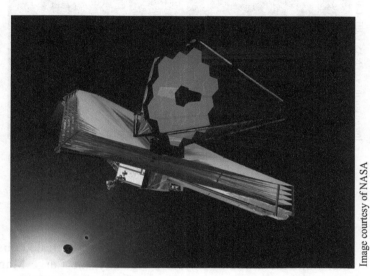

Image courtesy of NASA

Figure 4.28 Artist conception of the James Webb Space Telescope. When launched, the JWST will be the largest infrared telescope ever deployed in space and is expected to revolutionize astronomy and astrophysics.

Ultraviolet Astronomy

Just one year after Herschel discovered infrared radiation, Johann Ritter reasoned that if there was an invisible form of light just off the red end of the visible spectrum, there might also be a form off the violet end of the spectrum. His experiments led to the discovery of the radiation that is just above violet, "ultraviolet" rays.

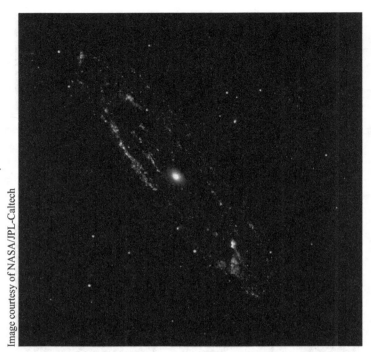

Image courtesy of NASA/JPL-Caltech

Figure 4.29 Image of the Andromeda galaxy taken by the GALEX spacecraft. Note that ultraviolet light easily traces out the spiral arm structure. This is due to the presence of hot massive stars along the arms.

Despite our concerns over sunburn and skin cancer, most ultraviolet radiation does not reach us (although the little that does is reason for caution). From longest to shortest wavelengths, we can categorize ultraviolet into uv-A, uv-B, and uv-C. *The Earth's atmosphere is nearly opaque to uv-B and uv-C radiation due mainly to the ozone layer. Thus almost all ultraviolet astronomical observations must be conducted from space. Since the 1970s there have been a number of ultraviolet telescopes placed in orbit around the Earth. The longest living of these was the *International Ultraviolet Explorer* (*IUE*) satellite. This spacecraft was launched in 1978 and ended its mission in 1996. The *IUE* was responsible for opening up the ultraviolet window to the universe and made many important discoveries related to stars, galaxies, and other important areas of astrophysics. The *IUE* is most famous for making very detailed observations of SN 1987A, a supernova that went off in the nearby dwarf galaxy, the Large Magellanic Cloud. This was the nearest, best-observed supernova since the beginning of the space program, and occurred before the *Hubble Space Telescope* was placed in orbit. The *IUE* became a workhorse in making space-based observations of the aftermath of the supernova explosion.

As a follow-up mission to the *IUE*, NASA launched the *Extreme Ultraviolet Explorer* (*EUVE*) satellite in 1992. This telescope was designed to observe ultraviolet wavelengths that were shorter than those observed by the *IUE*. The mission of the *EUVE* was completed in January 2001.

Image courtesy of NASA/JPL-Caltech/C. Martin (Caltech)/M. Seibert (OCIW)

Figure 4.30 GALEX image of the red giant star Mira in the constellation Cetus. Mira was the first star recognized to be a variable star in 1596 as its brightness changes by an impressive factor of 1,500 times between maximum and minimum brightness. This GALEX image shows evidence for shock-heated gas in a 10 light-year long trail of hydrogen coming from the star.

The *Galaxy Evolution Explorer* (*GALEX*) satellite was placed in space by NASA in April 2003. This spacecraft consists of a 50-cm diameter ultraviolet telescope, which is sensitive in the wavelength range from 135 nm to 280 nm. The primary goal of GALEX is to study other galaxies over enormous distances. In May 2012, the operation of the satellite was turned over to Caltech as a cost-saving exercise for NASA. GALEX was decommissioned in 2013.*

X-Ray and Gamma-Ray Astronomy

In 1895, Wilhelm Roentgen experimented with high-voltage cathode ray tubes, and accidentally discovered what he called x-rays. Years later, astronomers would discover that x-rays are also being emitted in space, and that they are associated with violent, high-energy events. It was the study of radioactive materials at the end of the nineteenth century that first led to the discovery of gamma-rays, and like all the other forms of electromagnetic radiation, astronomers found that there are also celestial sources of these dangerous rays.

*At wavelengths shorter than ultraviolet light, photons act more like particles than waves. Since the energy per photon is inversely proportional to its wavelength, x-ray and gamma-ray photons are highly energetic and can penetrate mirrors instead of being reflected using standard telescope designs. Since high-energy astrophysical sources are expected to emit copious amounts of high-energy photons, observing the cosmos at x-ray and gamma-ray wavelengths proves extremely valuable for studying these objects.

The trick to redirecting incoming x-ray photons so that they can be brought to a focus and create an image is the same method that is used to redirect a rock when you throw it at a lake. If the rock is thrown in such a way that it strikes the surface of the lake in a perpendicular direction, the rock will travel vertically down until it reaches the bottom of the lake. If, however, the rock is thrown such that it impacts the water's surface at a small angle with respect to the water, the rock will skip off the surface and travel a fairly long distance. This technique for redirecting a rock is known as grazing incidence and is used in the design of x-ray telescopes to change the path of incoming high-energy photons.

For x-ray telescopes, the primary mirror or lens found in optical telescopes is replaced by a series of nested mirrors (Figure 4.31). X-ray photons strike the surface of these mirrors at a small angle and get redirected toward an imaging camera or detector.

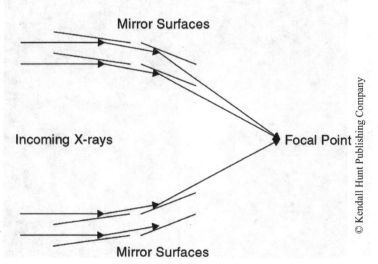

Figure 4.31 The grazing incidence design allows high-energy x-ray photons to be reflected from the surfaces of a series of mirrors so that they are directed toward a common focal point.

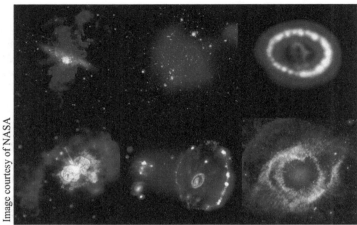

Image courtesy of NASA

Figure 4.32 A variety of objects observed by the Chandra X-ray Observatory.

Since the ozone layer in our atmosphere completely blocks all x-rays and gamma-rays from reaching the ground, the only way to observe the universe at these wavelengths is to place our detectors in space. There have been several x-ray satellites that have been launched into space by multiple countries. The most successful of these have been *Einstein*, *ROSAT*, *XMM-Newton*, and *Chandra*. Currently the largest x-ray telescopes in operation include *XMM-Newton* and *Chandra*. *XMM-Newton* is mainly a European-based mission, while *Chandra* is owned and operated by the United States. All of these x-ray satellites have opened up new frontiers in the exploration of high-energy phenomena, including black holes, galaxy clusters, and active stars.*

Launched in 2004, the *Swift* space telescope has multiple capabilities, including x-ray, ultraviolet and optical telescopes with the mission of helping astronomers to further study gamma-ray bursts (GRBs) and locate supernovae. A GRB is an incredibly brief, powerful explosion occurring far from Earth (thankfully). It is estimated that a 10-second GRB may release more energy than our Sun could in its theoretical lifetime. Astronomers believe a GRB can result from a supernova explosion, the formation of a black hole, a collision between pulsars or perhaps some other as yet unknown, exotic event. *Swift* picks up the burst of x-rays that accompanies a GRB and then points out the source and communicates with astronomers on the ground so that other telescopes can search the same part of the sky. *Swift* has conducted a variety of investigations in addition to studying GRBs, including the study of

Image courtesy of NASA/CXC/University of Toronto/ M. Durant et al.

Figure 4.33 Chandra image of the Vela pulsar. This neutron star is spinning at nearly 70 percent the speed of light as it completes 11 revolutions per second.

comets, stars with suspected exoplanets, neutron stars and black holes. In 2018, NASA announced that the mission was renamed the Neil Gehrels Swift Observatory, in honor of the principal investigator who died in 2017.

*Gamma-ray space telescopes–the most recent family of telescopes–use a different technique for recording photons than that used for x-ray telescopes. Since gamma-ray photons have extremely high energy, even grazing incidence does not help in trying to focus the incoming photons. Detectors for gamma-ray telescopes are essentially particle detectors, and are very similar to the detectors used in ground-based particle accelerators. The direction of incoming gamma-rays can be inferred by noting which end of the particle detector registers the photon first, thus rapid time resolution is critical for obtaining spatial information of incoming gamma-rays.

The *Compton Gamma Ray Observatory* (*CGRO*) was one of NASA's "Great Observatories," designed to detect gamma-rays (the other spacecraft in this series includes the *Hubble Space Telescope*, the *Chandra X-ray Observatory*, and the *Spitzer Space Telescope*). The *CGRO* was launched by the space shuttle *Atlantis* in 1991, and was de-orbited in 2000 when two of its three gyroscopes failed. Due to the increased risk of losing control of the massive spacecraft, NASA decided to make a controlled re-entry of the vehicle over the Pacific Ocean.

One of the main scientific achievements of the *CGRO* was the detailed monitoring of GRBs, the

5 arcminutes

Image courtesy of NASA

Figure 4.34 The bright spot in the center of the image is the 1,000th supernova detected by *Swift* back in 2015, observing x-ray, u.v., and visible light from the event.

Image courtesy of NASA E/PO, Sonoma State University, Aurore Simonnet

Figure 4.35 Artist conception of the Fermi Gamma-Ray Space Telescope in orbit around the Earth.

Image courtesy of NASA/*Fermi*

Figure 4.36 All-sky map of gamma-ray sources detected by Fermi during its first three years of operation. The plane of the disk of our Milky Way galaxy is clearly visible as the bright band across the middle of the image.

sudden short-term dramatic increase of gamma-rays coming from a specific direction on the sky. These objects were initially detected by military satellites in the 1960s and later disclosed to the scientific community two decades later. The *CGRO* helped to determine that these sources are not only associated with objects in our Milky Way galaxy, but are also detected in other galaxies throughout the universe. In 2008, NASA placed the *Fermi Gamma-Ray Observatory* into Earth's orbit. This telescope is the most powerful gamma-ray observatory ever placed in space and is designed to study GRBs, active galaxies, pulsars, dark matter, and other astrophysical objects. The *Fermi Gamma-Ray Observatory* is currently conducting major observations that will revolutionize our understanding of the high-energy universe at gamma-ray wavelengths.

Gravitational Wave Astronomy

On February 11, 2016, the Laser Interferometer Gravitational-Wave Observatory (LIGO) and the Virgo collaboration announced the first direct detection of gravitational waves. The existence of gravitational waves, ripples in space-time, was postulated by Albert Einstein as a consequence of his General Theory of Relativity. As objects accelerate they give rise to a disturbance in the curvature of space-time that propagates at the speed of light, c. The effect on space with the passage of a gravitational wave is to slightly squeeze space in one dimension while at the same time stretching space in a perpendicular direction.

The amount of spatial squeezing and stretching expected from gravitational waves emitted by astrophysical sources is on the average a fraction the size of a proton in the vicinity of the Earth. To detect such a small effect, the principle of the interference of light is used to monitor the combination of laser light as it traveled down two separate arms approximately 4 km in length. The light is reflected back toward the source by mirrors and recombined. The path length that the light travels along the two perpendicular arms of the interferometer can be adjusted so that the combined light after reflection undergoes destructive interference. When a gravitational wave passes by the detector, the slight change in the size of the spatial dimensions cause the path length that the light travels to change. This will destroy the perfect destructive interference of the combined light signals and thus signify that the light paths have changed in length.

The main problem in conducting a LIGO-like experiment on Earth is the elimination of false signals generated by earthly events. For example, earthquakes and tremors, and nearby passing trucks can shake the experiment to a degree that a "signal" will be recorded. Scientists and engineers have been working for several years trying to isolate the experiment for this type of noise. The result is the construction of advanced LIGO, which began operation in September 2015.

On September 14, 2015, a gravitational wave event was detected by the LIGO device in Louisiana and the state of Washington. Detailed analysis of the gravitational wave event indicates that it was probably due to the merging of two black holes with combined masses approximately sixty times the mass of our

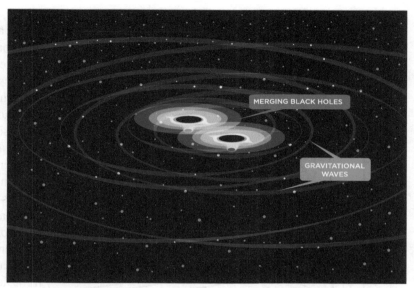

Figure 4.37 The merging of massive, compact objects such as black holes are expected to generate gravitational waves. The event detected on September 14, 2015 by LIGO is believed to have been caused by the merger of two black holes with masses of 36 and 29 times the mass of our Sun. (Artist's rendering)

Sun. This detection begins a new era in astronomy and astrophysics as different cosmic phenomena can now be studied via gravitational radiation rather than the traditional use of electromagnetic radiation.*

4.9 Spectra

**In order to study light in even more detail and learn more about its sources, astronomers use a device known as a *spectrometer* to separate light into components of different wavelengths. This is probably most familiar as visible white light being passed through a *prism* and separated into the colors of the visible spectrum. The device that separates the light in a spectrometer is a piece of glass with many lines, very close together, etched into it, called a *diffraction grating*. Diffraction is the spreading and interference of light rays caused when they encounter an obstacle. The lines in a grating are so close together that they provide obstacles for the light waves. Different wavelengths are diffracted by different amounts, so when passing through a diffraction grating, light gets separated into the different wavelength components of its spectrum.

Temperature

One of the first useful things astronomers can learn from a spectrum is the *temperature* of the object that it came from. Temperature is actually a measurement of energy. If the temperature in a room is

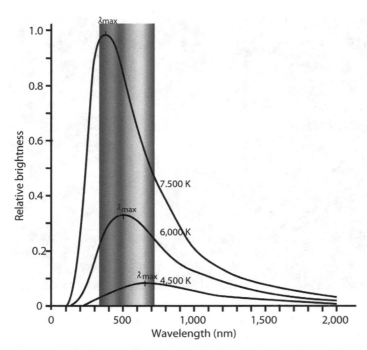

Figure 4.38 Spectra for a blue-7500 K, yellow-6000 K, and red-4500 K star. Note the hottest, blue, star's brightness peaking at a shorter wavelength and, therefore, higher frequency. Note the opposite for the coolest, red, star.

higher, it means that the air molecules are moving faster and are, therefore, more energetic. If a room is cooler, the air molecules are slower and less energetic. The higher the frequency of an electromagnetic wave, the more energetic it is. This is because the waves repeat their cycle more rapidly. More energetic electromagnetic waves mean a more energetic, higher temperature source. So, objects that give off high-frequency gamma rays or x-rays are very hot, and objects that give off only low-frequency radio waves are very cold.

Stars that give off mostly red light are cooler, while stars that give off mostly blue light are hotter. Yellow stars are in between. Stars that give off light across most of the visible spectrum in about the same amounts will appear white. These stars are hotter than yellow stars but are cooler than blue stars. Stars that appear blue give off most of their light at invisible ultraviolet wavelengths, but they give off more blue light than any other visible color.

The reason for being careful to say that a star gives off "mostly" red or "mostly" yellow light is that stars actually give off many wavelengths of light, but the wavelengths, and, therefore, the colors that we see, are those that they give off the most. It can be seen in Figure 4.38 that the spectrum of a star will peak at a specific wavelength. This determines the color the star will appear to be and will tell astronomers its exact temperature.

Composition

Another important property that a star's spectrum can reveal to astronomers is its composition. There are three kinds of spectra, as shown in Figure 4.39. A *continuous spectrum* comes from a hot, dense object, such as a light-bulb filament, or a gas under very high pressure, like the core of a star. A *bright-line* or *emission spectrum* comes from a hot gas, and a *dark-line* or *absorption spectrum* is what astronomers see when they observe stars.

It is not difficult to see in Figure 4.39 that subtracting the emission line spectrum from the continuous spectrum will leave the absorption spectrum. This is exactly what happens in stars. If light could come to us directly from the core of a star, since the gas in the core is under tremendous pressure from the mass of the star's outer layers, we would see a continuous spectrum. If we could look only at the light from the star's outer layers that are heated by the energy from the star's interior, we would see an emission spectrum. But what we see is what remains after light from the core of the star has passed through the star's outer layers—an absorption spectrum.

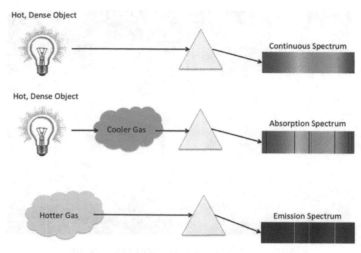

Figure 4.39 The three types of spectra.

The reason that these spectra tell us the composition of stars is that the emission spectrum of a particular gas is like a fingerprint or DNA pattern. No two gases have the same spectrum. Danish physicist Niels Bohr determined that the spectrum of a gas depends on the structure of the atoms that make up the gas.

When an atom absorbs energy, it will give back, or emit, the energy in the form of particles of light called *photons*. The exact amount of energy, and, therefore, the frequencies and wavelengths of the photons the atoms will emit, is dependent on the structure of the individual atoms. Atoms will absorb the same amounts of energy they emit, so when energy comes from the core of a star, it is absorbed in the outer layers, and what is left for astronomers to see is a dark-line or absorption spectrum. The wavelengths of the spectral lines that are absorbed can be matched with the bright lines of different emission spectra to determine the exact composition of a star. Stars, as it turns out, are made mostly of hydrogen and helium with smaller amounts of many other types of gases.** In fact, helium was first discovered by studying spectral lines from the Sun, and the element was named accordingly ("helios" = sun).

Figure 4.40 Bohr.

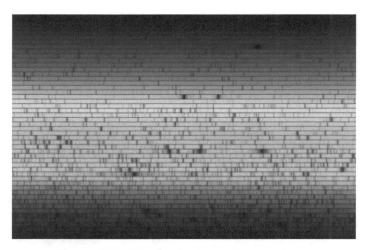

Figure 4.41 The Sun's absorption spectrum.

Years later, helium was also discovered on Earth.

Fortunately for astronomers it was discovered by Annie Jump Cannon, of the then Harvard College Observatory, that there are only seven basic types of stellar spectra. There are numbered sub-types of each *spectral type* or *class*. For example, the Sun is a "G-2" star. It is hotter than a G-3, and cooler than a G-1. The letters *O*, *B*, *A*, *F*, *G*, *K*, *M* are in order of decreasing temperature. A simple way to remember this is "B" for blue and then recalling the star colors in order of decreasing temperature. B-type stars are very hot blue stars. A-type stars are hot and white. G-types are cooler yellow stars like our Sun, and M-type stars are the coolest, red stars.

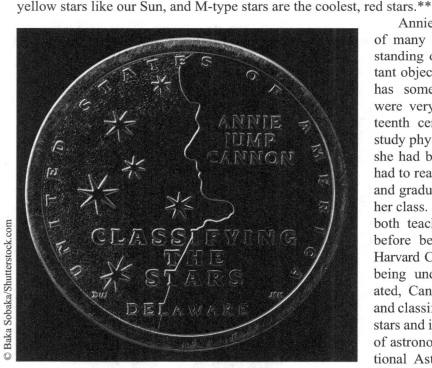

© Baka Sobaka/Shutterstock.com

Figure 4.42 Annie Jump Cannon has been recognized by the U.S. Mint in their new one-dollar coin "American Innovation" series.

Annie Jump Cannon is just one of many contributors to our understanding of what the light from distant objects can tell us, but her story has some unique qualities. There were very few women in the nineteenth century going to college to study physics. In the case of Cannon, she had become deaf as a child and had to read lips, yet she still excelled and graduated as the valedictorian of her class. Cannon eventually pursued both teaching and graduate studies before being hired to work at the Harvard College Observatory. Despite being underpaid and underappreciated, Cannon went on to catalogue and classify hundreds of thousands of stars and influence future generations of astronomers. In 1922, the International Astronomical Union adopted Cannon's method as the official spectral classification system. Cannon was the first woman to ever receive an honorary degree from Oxford, and the first woman to ever receive the Henry Draper Medal of Honor from the National Academy of Sciences.

Thanks to the work of pioneers like Annie Jump Cannon, spectroscopy is used widely for such large projects as sky "surveys." One such project is the Sloan Digital Sky Survey, which began analyzing and cataloging stars and galaxies in 2000. The SDSS continues its work today, having imaged many millions of objects to help create a thorough map of our galaxy and beyond. The data collected by the SDSS is made available to the public, and you can find it on the SDSS website.

The Solar System— Part 1

> *"God has constructed the universe in a truly marvelous way. As we study it, the universe continually surprises and delights us by challenging our understanding of how things work. Our solar system is a great example of this. We can actually see much of the solar system on a cloudless night… Every new discovery in astronomy is a surprising and delightful revelation that God is even more amazing, creative, and powerful than we previously supposed."*
>
> **—astrophysicist Jason Lisle**

5.1 A Survey of the Solar System

*What is contained in the solar system? Of course, the Sun, which contains more than 99.9 percent of the solar system's mass, is in the solar system. But since the Sun is a star, we will defer discussion of it to Chapter 8. We already mentioned that there are eight planets, but what is a planet? A planet is a large body that orbits the Sun, and that has "cleared" its orbital neighborhood of other objects (sorry, Pluto!).

Figure 5.1 This rendering of our planets is not drawn to scale.

© Baka SobakaMaxx-Studio/ Shutterstock.com

Millions of individual objects probably orbit the Sun, but only eight are large enough to be called planets. We shall return later to the question of the minimum size for a planet. We ought to mention that the Sun's gravity compels planets and other objects to orbit the Sun. Orbits generally are elliptical,

---

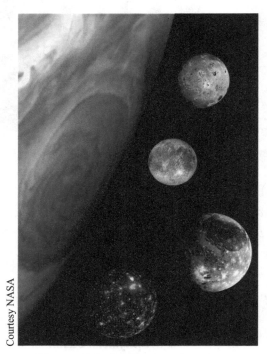

Courtesy NASA

Figure 5.2 Jupiter has four large satellites, shown here to scale with Jupiter.

Courtesy NASA

Figure 5.3 Saturn's satellite Titan, the second largest satellite in the solar system, compared in size to Earth and Moon.

but planetary orbits are very close to being circular. The orbits of the planets lie in nearly the same plane, so the Sun's planetary system is very flat. Earth's orbital plane, the ecliptic, is the plane that we use for reference.

A satellite is a body that orbits a planet (Figure 5.2). Earth has one satellite, the Moon. The Moon is the name of Earth's natural satellite. Sometimes people call satellites of other planets moons, but technically, this is not proper. While the correct term is *satellite*, there is no escaping the use of "moon" as well. Two planets, Mercury and Venus, have no satellites, but other planets have scores of satellites. Two satellites, one of Jupiter's and one of Saturn's, are larger than the planet Mercury (Figure 5.3). You may wonder why a satellite larger than Mercury is not a planet too. Size is not the issue here; it is what an object orbits that matters. If these two satellites orbited the Sun in their own right, they would be planets, but since they orbit planets, they are satellites.

What are the millions of smaller objects that orbit the Sun? The smaller objects in the solar system generally are either **minor planets** (asteroids) or **comets**. For many years, the main distinction between minor planets and comets was composition. Astronomers thought that minor planets were rocky, while comets were made of various ices, such as water, dry ice, and frozen methane. However, in recent years there has been a blurring of this distinction, for many asteroids have turned out to be icy too. This difference in composition still is important, but perhaps more important is the different sorts of orbits that comets and minor planets have. Minor planets have orbits that are similar to planetary orbits. That is, their orbits are nearly circular and lie very close to the plane of the orbits of the planets (Figure 5.4). The common name for minor planets is asteroids, but astronomers prefer to call them minor planets, because it emphasizes their orbital similarity to planets. On the other hand,

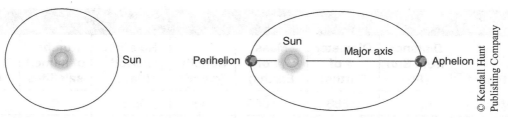

Figure 5.4 A comparison of a circular orbit and a very eccentric orbit.

comets tend to have orbits very different from planets. Their orbits are very eccentric ellipses. This causes a comet to spend very little time close to the Sun (a point we call **perihelion**) and much of its time far from the Sun where it is very cold. Another difference is that cometary orbits generally do not lie close to the plane of the orbits of the planets, but instead are highly inclined to the solar system plane (Figure 5.5). Planets and minor planets orbit the Sun in the same direction, counterclockwise as viewed from above Earth's North Pole. While many comets orbit in this direction, many orbit **retrograde**, or clockwise, as viewed from above the North Pole. To many astronomers, the vast differences in orbits and composition between comets and minor planets suggest different origins and histories.

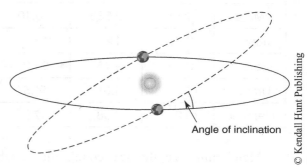

Figure 5.5 The orbit (dashed curve) is shown inclined to the plane of the solar system (solid curve). Here it is exaggerated for effect.

Two Types of Planets

Table 5.1 lists a number of orbital and physical characteristics of the planets. Many of the characteristics are expressed in terms of Earth's traits. For instance, the second column lists the average distance of each planet from the Sun in terms of Earths' distance from the Sun. We call the average Earth–Sun distance the **astronomical unit**, usually abbreviated AU. Notice that the first four planets, Mercury, Venus, Earth, and Mars, are closely spaced, but that the next four planets, Jupiter, Saturn, Uranus, and Neptune, are much more widely spaced. In fact, the first large break is between Mars and Jupiter. This break is obvious in many other ways too, and we shall use this break to make a distinction between two types of planets. The first four planets are similar to one another, and since Earth is in this group, we call these the **terrestrial planets**, or Earth-like, planets. Earth is the largest of the terrestrial planets. Similarly, the last four planets are very similar, and Jupiter is the largest of those planets, so we call them the **Jovian planets**, or Jupiter-like, planets.

Planet	Distance from Sun (AU)	Diameter (# of Earths)	Mass (# of Earths)	Density (g/cm³)	Rotation period (days)	Number of named satellites	Rings
Mercury	0.387	0.383	0.055	5.43	58.6	0	No
Venus	0.723	0.949	0.815	5.24	−243	0	No
Earth	1.000	1.000	1.000	5.515	0.997	1	No
Mars	1.52	0.533	0.107	3.94	1.03	2	No
Jupiter	5.20	11.2	318	1.33	0.414	70+	Yes
Saturn	9.58	9.45	95.2	0.70	0.444	60+	Yes
Uranus	19.3	4.01	14.5	1.30	−0.718	27	Yes
Neptune	30.2	3.88	17.2	1.76	0.671	14	Yes

Table 5.1   Properties of the Planets

Mentioning size, the next column in the table gives the diameter of each planet in terms of Earth's diameter. Of course, in these units Earth's size is one; notice that the other three terrestrial planets are smaller than Earth. On the other hand, you can see that the four Jovian planets are far larger than any terrestrial planet. The next column compares mass, with Earth being the standard of measure once again. You can see that, as with size, Earth is the most massive terrestrial planet, but that all the Jovian planets dwarf Earth's mass. Knowing a planet's mass and, size, we can compute its density. The next column of Table 5.1 lists the density of each planet. Density is a measure of how much matter occupies a certain volume. Density depends upon composition, so density is very important in deducing what a planet's composition likely is. As you can see, the densities of the terrestrial planets range from 3.9 to 5.5 g/cm³. This is consistent with rocky composition, so we sometimes call the terrestrial planets the rocky planets. Now look at the densities of the Jovian planets. Their densities are less than 2 g/cm³ and are far less than those of the terrestrial planets. Since rocks normally have density of at least 3 g/cm³, it is clear that the Jovian planets cannot contain much rocky material. Then what is the composition of Jovian planets? They are made largely

Courtesy David Rives

Figure 5.6 Larger spot near the top is Venus, caught here in a rare transit across the Sun's face. The other spots are sunspots.

of hydrogen and helium. On Earth's surface, these elements are gases, so we frequently call the Jovian planets the gas giant planets. The Jovian planets have atmospheres, and the hydrogen and helium there are gaseous, but within their vast interiors, the pressure is so great that most of the hydrogen and helium there are liquid.

The next column lists the rotation periods of the planets, either in hours or in days. Venus and Uranus have negative signs; this means that these two planets rotate backward, or retrograde. You can see that two terrestrial planets, Mercury and Venus, rotate very slowly. Mars and Earth rotate much more quickly, with either one taking about twenty-four hours to rotate. However, notice that the Jovian planets all rotate very quickly, more quickly than Earth, the terrestrial planet with the shortest rotation period. This suggests another difference between the two types of planets—Jovian planets have shorter rotation periods, while terrestrial planets have longer rotation periods. The next column compares the number of satellites. There are only three satellites among the terrestrial planets, but the satellites of the Jovian planets total nearly 200, with the discovery of new ones continuing, especially in the cases of Jupiter and Saturn. The last column lists whether a planet has a ring system. All the Jovian planets have ring systems, but none of the terrestrial planets do, so this appears to be a distinction between the two types of planets. Table 5.2 summarizes the differences between the two types of planets.*

Table 5.2 Comparison of the Terrestrial and Jovian Planets							
Planet Type	Distance from Sun	Diameter	Mass	Density	Rotation period	Number of satellites	Rings
Terrestrial	Near	Small	Small	High	Long	Few	No
Jovian	Far	Large	Large	Low	Short	Many	Yes

The Origin of the Solar System

There are two competing models for the origin of our solar system. This is a pretty straightforward issue; either the solar system was created, or it wasn't. It's impossible to use science to "prove" either position conclusively. All we can do is see which scenario fits the evidence the best. Attempting to explain a one-time event from the past creates a different set of challenges than those of operational science. Historical science by its very nature will be far more speculative and will rely heavily on a priori philosophical commitments.

The creation model begins with the historical record from the Bible. The Bible says in the beginning *God created* the heavens and the Earth. The Bible doesn't give us a specific date for this event but from other passages in the Bible we can calculate that this would have been about 6,000 years ago. Biblical creation models begin with the authority and historicity of scripture, and then interpret evidence accordingly.

On the other hand, evolutionary models begin with naturalism. If the underlying assumption is that all there has ever been is natural law at work in the universe, and that there is no such thing as the supernatural, then it follows that creation must be ruled out and any origin story must rely heavily on the belief that given enough time, matter and energy can create everything.

There have been several secular theories about how the solar system came to be. In the early 1900s, when Einstein was publishing *special relativity*, the **planetesimal theory was being advanced by American geologist Thomas Chamberlain: he suggested that the gravity of a passing star pulled gas and heavier elements from the Sun, which condensed to form planets. Pretty much nobody subscribes to this theory today—the distance between stars is so great that near-collisions like this are unlikely, and the compositions of the inner planets are not like the Sun's. In 1919, Sir James Jeans proposed the **tidal theory**, where hot gas blew off the Sun and made planets, which were first gaseous, then liquid, and finally solid. Again, the Sun's composition and the compositions of the inner planets are very different, and as there is no other evidence to support it, no one accepts this theory today. The now discredited **double star theory** of the 1930s said that the Sun was once a binary star, but that the companion star blew up and the debris cloud was the material from which the planets formed. Again, there's no evidence for such a cataclysmic event. The **condensation theory**, which was developed in the 1950s, contains the rudiments of the **solar nebula theory** most astronomers propose today: the material from an exploded star or stars collected into a nebula out of which the Sun and planets formed.**

The Solar Nebula Theory (Nebular Theory)

The secular consensus today is that the solar system began about 4.6 billion years ago. This age is based on a series of assumptions. This is historical science rather than observational. The formation scenario goes about like this:

© Triff/Shutterstock.com

Figure 5.7 Artist's rendering of our own solar nebula.

There was a great cloud in space—a nebula. This nebula was made of hydrogen and helium gas, as well as small amounts of all the other elements familiar to us today. The cloud was trillions of miles in diameter, and yet, it was remarkably thin and tenuous. Left to itself, probably nothing would have become of the cloud—it might have simply drifted outward and dissipated. But the nebula was not left to itself. A powerful **shock wave thundered through this region of space. It may have been a gravitational density wave passing through our galaxy, or perhaps it was the explosion of a nearby star—a supernova. The shock wave compressed the cloud, and as it collapsed, the cloud's turbulence was transformed into a slow, grand pirouette.

As it shrank, the cloud's spin accelerated. Gas molecules struck each other, creating heat. Soon the temperature at the heart of the cloud reached a staggering 15 million degrees Fahrenheit—hot enough to fuse hydrogen atoms into helium. With the creation of this new element, a small amount of atomic binding energy was released—nuclear fusion began, and the super-dense ball of gas at the center of the cloud began to shine—a star was born. Even as our

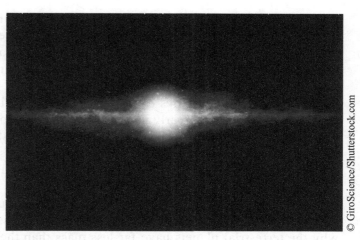

Figure 5.8 Proposed birth of the Sun, the solar nebula becomes a revolving disk of matter.

Sun formed, leftover gas and dust continued swirling, gradually flattening into a **disk**. As gas condensed and solid particles collected and stuck together (a process called accretion), planetesmals, the building blocks of the planets, were created. The planetesimals grew like rolling snowballs, sweeping up material in their paths. Marble-sized bits of matter soon became mountain-sized. And still the planetesimals grew, becoming **protoplanets**, still molten, but of roughly the same size as they currently are. The densest and hottest part of the nebula was at the center—here the planets that formed were hard, solid, metallic, rocky—the Earth-like, or terrestrial worlds. Farther out, where temperatures were cooler, ices condensed, creating great **volatile** gas giants—the Jupiter-like, or Jovian worlds.**

So, scientists committed to evolution think that the solar system formed from a large cloud of gas and dust that collapsed under its own gravity. Gas clouds do not spontaneously collapse like this, so the process supposedly began by some unknown mechanism—the fortuitous "shock wave." The shock wave argument becomes circular. How do we know there was a shock wave? Because the solar system is here! *According to this theory, most of the gas fell to the center to form the Sun. The small percentage of remaining material flattened into a disk, from which the planets slowly formed. The first step in this process of planetary formation was for the small dust particles in the

Figure 5.9 The theory claims that planetesimals grow to become protoplanets thanks to accretion.

flattened disk to stick together. How this happened is unknown too. Gradually, the small particles grew larger by combining. Astronomers call these hypothetical particles **planetesimals**. Eventually, some planetesimals grew large enough so that their gravity began to dominate within certain regions of the solar system. These dominant planetesimals became the seeds to form the planets.

Why are there two types of planets, the terrestrial and Jovian planets? According to the evolutionary theory, the early sun heated up and evaporated all the lighter material (mostly hydrogen) from the planetesimals in the inner solar system, close to the Sun. Since those planetesimals lost the lighter material, the composition was rocky, so the planets that formed from them (the terrestrial planets) have rocky composition. Planetesimals farther from the Sun kept their lighter material, so the planets that formed from them (the Jovian planets) have lighter composition. The lighter material accounted for the bulk of the mass, so when the planetesimals closer to the Sun lost the light material, they lost most of their mass. This accounts for why the terrestrial planets have far less mass than the Jovian planets. While this theory can explain the differences in mass and composition between the Jovian and terrestrial planets, it cannot explain other things. For instance, it does not explain why Jovian planets rotate more rapidly than terrestrial planets.* Moreover, the model violates a fundamental rule of physics called conservation of angular momentum. Based on the nebular theory, the Sun should have a tremendous amount of angular momentum, spinning far faster than it does. Instead we observe too little angular momentum for the Sun and a disproportionate amount for the planets.

It turns out that you cannot take the process of forming clumps of dust and simply extrapolate that into forming planets. We know from experience that dust particles will stick together. Experimental evidence shows that clumps of dust can form in the vacuum of space as well. However, those clumps reach a critical size and then they do not grow together any further. In the scenario proposed by the solar nebula model, these pebble-sized objects begin to impact each other at such high speeds that instead of sticking together they break apart in the collisions. Gravity cannot overcome this effect unless accretion produces much larger objects, but there is no evidence of how the process would ever get objects to reach that size. Astronomers may refer to computer simulations to support their claims, but we see a lack of evidence in the real world to support the notion of going from dust clumps to planets.

Evolutionary astronomers know there are problems with the solar nebula theory. That's why we see quotes like this one in astronomy textbooks: "once these planetesimals have been formed, further growth of planets may occur through the gravitational accretion into large bodies but just *how that takes place is not understood*". And in the scientific journals we see comments like this: "the formation of planetesimals [small asteroids] the kilometer-sized planetary precursors is still a *puzzling process.*"

Another problem that secular researchers are aware of is something called "the Jovian Problem." We will return to this problem when discussing each of the gas giants later. The problem is that based on the solar nebula theory, the four largest planets either should not have formed at all, or should have formed in very different locations and in very different time frames. In other words, what we observe does not fit with what the model predicts. That is not to say astronomers are not working to "fix" the model, but it is a serious problem.

Likewise, the study of exoplanets (planets orbiting distant stars) has produced observations that conflict with the solar nebular theory, which is supposed to work for any solar system, not just our own. For example, when multiple planets are found in a system, they tend to be like "peas in a pod"—all the same size and having similar physical characteristics. Contrast that to the variety among our eight planets. Exoplanet research also finds "hot Jupiters" and "hot Neptunes"—large planets like our own gas giants that are far closer to their star than the nebular theory says would be possible. Also, according to the nebular model, planets should orbit in the same direction—clockwise or counterclockwise—that their star is spinning. Furthermore, they should orbit the star within the plane of the star's equator. However, some exoplanets are observed to orbit backward, and others have orbits that are highly inclined to their star's equatorial plane.

Finally, the study of other disks that are supposedly in the early stages of solar system formation shows that there is not enough material in these disks to produce exoplanets. Astronomers used Chile's Atacama Large Millimeter Array (ALMA) radio telescope to study hundreds of star systems that they considered to be "young," surrounded by dust-particle disks. They then compared the masses of these disks to the total masses of exoplanets around what they considered to be "older" stars of similar size. They found that the disk masses were much less than the total exoplanet masses, sometimes by as much as a factor of 100. In other words, secular scientists are claiming that exoplanets form from the material in these thin disks, but the disks simply don't have enough material in them to actually *make* the exoplanets! As *Scientific American* said in 2011, "Observers in any field of science take a peculiar pleasure in seeing their theorist colleagues collapse into sobbing heaps, but it happens with unnerving regularity with exoplanets. Modelers have consistently failed to predict the diversity of planetary systems out there." In other words, the solar nebula model is not working. What the model predicts is not what is being observed.

Demonstrating that the current consensus should be viewed with skepticism is admittedly not the same as proving that the creation model is true. However, as much as the evidence should raise doubts about the secular model, the underlying worldview should be doubly concerning to the Christian wrestling with whether or not to "trust the science." As evolutionist Richard Lewontin wrote,

> "Our willingness to accept scientific claims that are against common sense is the key to an understanding of the real struggle between science and the supernatural. We take the side of science in spite of the patent absurdity of some of its constructs, in spite of its failure to fulfill many of its extravagant promises of health and life, in spite of the tolerance of the scientific community for unsubstantiated just-so stories, because we have a prior commitment, a commitment to materialism.
>
> It is not that the methods and institutions of science somehow compel us to accept a material explanation of the phenomenal world, but, on the contrary,

that we are forced by our a priori adherence to material causes to create an apparatus of investigation and a set of concepts that produce material explanations, no matter how counter-intuitive, no matter how mystifying to the uninitiated.

Moreover, that materialism is absolute, for we cannot allow a Divine Foot in the door."

What Lewontin is saying is consistent with what appeared in Chapter 1. Science does *not* require anyone to embrace materialism (naturalism), but if you do embrace materialism then you can never "allow a Divine Foot in the door." That is easy for a materialist, as they do not believe the Divine exists.

Table 5.3 More Properties of Planets	Mercury	Venus	Earth	Mars	Jupiter	Saturn	Uranus	Neptune
Magneto-sphere	Weak	None	Present	None	Strong	Strong	Present	Present
Compared to Earth	0.01	0	1	0	20,000	600	50	25
Orbital Period	0.24 y	0.62 y	1 y	1.9 y	11.9 y	29.5 y	84 y	165 y
Orbital Eccentric-ity	0.21	0.007	0.02	0.09	0.05	0.06	0.05	0.009
Axial "tilt"	0°	177°	23.5°	25°	3°	26.5°	98°	29.5°
Seasons	No	No	Yes	Yes	No	Yes	Extreme	Yes
Albedo	6%	76%	39%	16%	51%	50%	66%	62%
Gravity	0.4 g 3.7 m/s²	0.9 g 8.8 m/s²	1 g 9.8 m/s²	0.4 g 3.7 m/s²	2.5 g 24.7 m/s²	1.1 g 10.4 m/s²	0.9 g 8.8 m/s²	1.1 g 11.1 m/s²
Primary Atmo-spheric Gases	Negligible	CO_2	N_2, O_2	CO_2	H_2, He, NH_3, CH_4			
Atmo-spheric Pressure (atm)	Negligible	90	1	0.01	Crushing			

5.2 Earth as a Planet

*It is helpful to begin our study of the other planets with Earth as a model, for some of the things that we learned about Earth may apply to those other planets. For instance, we know that Earth is differentiated into a core, mantle, and crust. There is evidence that other planets are at least partially

Figure 5.10

differentiated as well. Many of the minerals found on Earth are also found in astronomical bodies. Meteorology is the study of Earth's atmosphere. Most planets have atmospheres too, though their atmospheres are very different from Earth's atmosphere. One of the best ways to explore the planets is to employ **comparative planetology**, the study of planets by noting their similarities and differences. In this context, Earth often serves as the standard for comparison. Since we know so much about Earth's Moon, it is a good basis of comparison to the satellites of the other planets.*

Tectonic Activity on Earth

Earth is the biggest, most massive terrestrial planet, and its interior is apparently still molten. Much of what we believe about the internal structure and behavior of the Earth comes from the study of earthquake waves. These waves have different properties and behave differently depending on the temperature, density, and physical state of the materials they travel through. These waves provide scientists with information about the different regions and boundaries within the Earth, leading to our current model—**beneath the crust is a **mantle**; beneath that is a fluid **outer core**, and at the center of it all is a solid **inner core**. Heat from the mantle works its way up to the surface, with amazing results: **volcanic eruptions**, **earthquakes**, and most impressive of all, a constant shifting of the Earth's crust in a process called **plate tectonics**. (from the Greek, "tektos," meaning "builder" or "roof"; the crust is our planet's roof.)** Earth is the only planet in the solar system that shows clear evidence of plate tectonics.

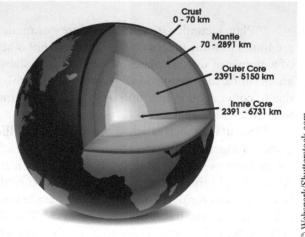

Figure 5.11 Cutaway view of the Earth.

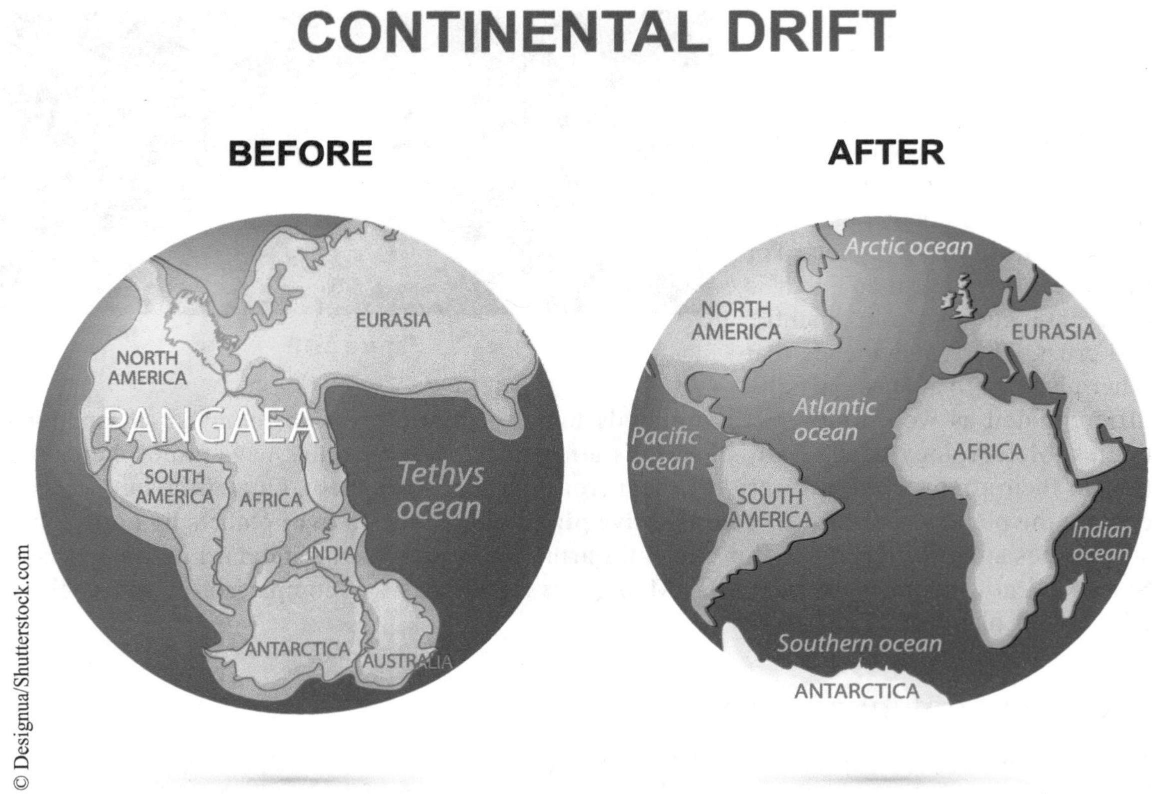

CONTINENTAL DRIFT

BEFORE   **AFTER**

Figure 5.12 Pangaea is the hypothetical "super continent" of the past; since then, the continents have moved apart.

Plate tectonics is a unifying theory or model which explains why earthquakes and volcanoes are found in some places on the Earth and not others. The theory is that the Earth's crust is made up of a couple of dozen "plates,"—actually big slabs of crust and some underlying mantle that move about (think of tomato soup brought to a low boil, with chunks of tomato soup "skin" drifting about due to convection in the fluid soup). This movement of the Earth's crust has caused the continents to drift.

Secular models suggest that over 200 million years ago, all of our planet's continents were gathered together into one great, "supercontinent" known as **Pangaea. Then came the breakup of the continents—North and South America split from Europe/Asia and Africa, and Antarctica and Australia split from all of them. One split occurred along what is now the mid-Atlantic ridge, where new crust was formed from underlying mantle basalt. And it's not just the continents in motion; the entire crust, ocean basins too, is moving.** The 200 million years is based on an assumption called uniformitarianism, which claims that whatever we see happening today can be extrapolated to tell us about the past ("the present is the key to the past"). The problem is that there is no scientific justification for uniformitarianism—it is simply an assumption that is accepted by many naturalists. In this case, since we observe very small, slow movements of the Earth's plates in a given year, the assumption is made that it took 200 million years to get from Pangaea to the present positions of the plates. Creationists like geophysicist John Baumgardner have developed scientific models of Pangaea to suggest catastrophic plate movements associated with a global flood—large movements in a short period of time.

Wherever crustal plates form and pull away from each other (divergent plate boundaries**), or wherever plates run into each other (**convergent plate boundaries**), earthquakes and volcanoes occur. At the divergent mid-Atlantic ridge, new land masses form, such as Iceland. Along places where oceanic plates meet and one or both plunge downward and melt (a process called subduction), great trenches form (the Marianas trench in the western Pacific); or in the case where continental plates collide and buckle upward, great mountain chains form (the Himalayas of India and Nepal). In some regions plates slide past each other, creating **transform** or **strike-slip** faults such as the San Andreas of California. To sum up, most major earthquakes and volcanoes happen principally along plate margins—unless there's a geologic "hot spot" under a plate, such as has given rise to the Hawaiian Islands in the Pacific Ocean.** Earthquakes can happen almost anywhere, but they are greatest in number and severity along plate boundaries.

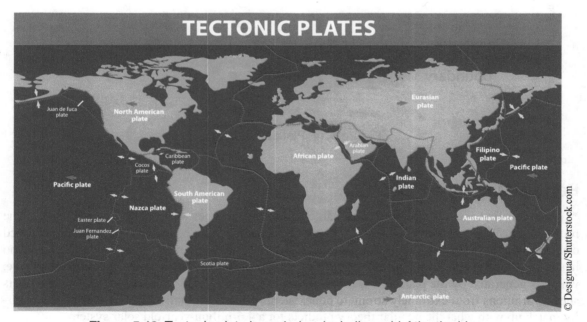

Figure 5.13 Tectonic plate boundaries, including mid-Atlantic ridge.

THREE TYPES OF PLATE BOUNDARY

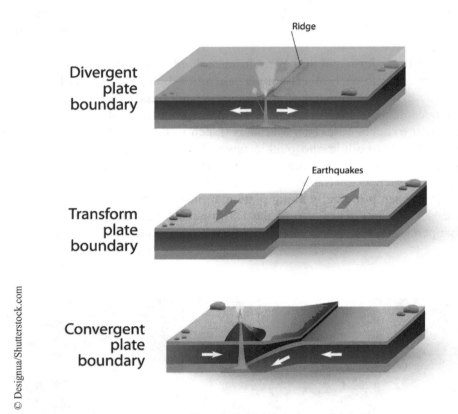

Figure 5.14 Plate boundaries.

Earth's Magnetic Field

When the first rockets equipped with cameras left the surface, we got our first real look at the round Earth. When Explorer 1 made it into space, instruments on board detected a highly charged electromagnetic field surrounding our planet: the **Van Allen belts (named for the rocket scientist who headed up the research). These particles were trapped in place by the Earth's powerful **magnetosphere**, which is generated by the planet.** When it comes to magnetic fields, a survey of the eight planets shows variations from extremely strong to nonexistent. What we actually observe does not necessarily match up with what the nebular theory predicts for a 4.6-billion-year-old solar system.

Earth's magnetic field causes all sorts of problems for evolution. Our magnetic field looks sort of like the field that would be generated if there was a giant bar magnet inside the Earth. Magnetic

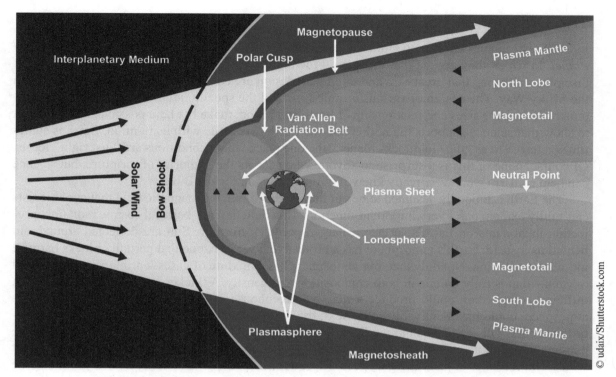

Figure 5.15 The Van Allen radiation belts and Earth's magnetosphere.

fields are supposedly generated by fluid motions inside the planets, an idea called the dynamo theory. The model is incomplete at best, and fails with some planets. It's hard to make it work for the Earth, even though the Earth is the planet it was invented to explain. The American Geophysical Union website quotes a study that says this: "the mechanism for generating the geomagnetic field remains one of the central unsolved problems in geoscience." The idealized mechanism for creating the dynamo effect suggests that the combination of a planet's rotation and hot liquid metal at its core generates a magnetic field, but it proves to be more complicated and difficult to explain.

An even bigger problem is that our magnetic field is decaying. For as long as we've been measuring it, its energy has fallen continuously. It loses half its energy about every 1,400 years. In this case we are not referring to "reversals" in the magnetic field (a different topic). We're talking about all the energy in the field overall and that energy is fading fast. Since it's getting weaker over time this means it was stronger in the past. If we apply secular, uniformitarian logic, traveling backward in time the field would double its energy every 1,400 years. There's a definite limit on how strong it could possibly have been. It turns out that the Earth's magnetic field could only be tens of thousands of years old at the most, not billions. Once again, creationists have developed scientific models that suggest non-uniform changes during the Genesis flood. Evolutionary dynamo theories fail to explain what we observe. That's why one evolutionist recently wrote: "magnetism is almost as much of a puzzle now as it was when William Gilbert wrote his classic text in 1600." The evidence suggests that Earth's magnetic field is "young."

The Uniqueness of Earth

In 2004, astronomer Guillermo Gonzalez and philosopher Jay W. Richards wrote *The Privileged Planet*, in which they argued that there is a growing body of scientific evidence from a range of disciplines showing that great care and intelligent design are conspicuous in our placement here in the Milky Way. The design of various cosmic laws and the specific architecture of our solar system, including the sizes and relationship of Earth and Moon, make life here possible. They attempt to show that science supports the notion that the Earth was made with us in mind. Earth is sometimes referred to as "the Goldilocks planet," the one place where conditions are "just right" for us. Of course, none of this is a surprise to anyone who accepts the authority of scripture, but science can help confirm what we already knew if our starting point is the Bible.

The Earth is uniquely designed to be our home. First of all, the magnetic field that evolutionists can't fully explain has a very important function. This field deflects harmful cosmic radiation, protecting inhabitants on Earth's surface. Charged particles from the Sun are deflected around and behind us instead. Earth's atmosphere has a protective layer of ozone that partially blocks harmful ultraviolet radiation. Not only that, our atmosphere has a mixture of gases—primarily nitrogen and oxygen—that is truly unique in both composition and density. A thicker atmosphere or one with a different composition could produce a massive greenhouse effect like that of Venus. A thinner atmosphere could mean unlivable cold. Next, the Earth rotates on its axis once about every twenty-four hours. This again is a unique feature for us. If the Earth rotated too slowly there would be extreme temperature changes on our planet. If it rotated too quickly, we would experience violent winds. But it rotates at just the right speed. Also, our planet's axis is tilted by 23.5 degrees and the Earth's orbit is nearly circular. This gives us moderate seasons and a relatively stable climate (despite some extreme views to the contrary).

There's another very important feature of our home planet. Seventy percent of its surface is covered in water. It's hard to comprehend how much water is here. The ocean basins are miles deep in some places. If you could raise the ocean bottoms up and bring all the mountains down so that the entire Earth's surface was leveled, the water in the oceans would spread out to cover the entire Earth, to a depth of well over a mile. There's a tremendous amount of water here, more than enough to flood the entire planet if only the Earth's surface was a bit smoother.

This water, enough to flood the planet, is one of the reasons we see Earth's surface continually going through changes that we refer to as erosion. Along with erosion from water, our atmosphere is responsible for wind erosion. The other major force behind the changes we observe has already been mentioned, namely plate tectonics. We will see in subsequent topics, that this combination of surface changes is unique to the Earth.

All this water is vital to life but its presence and abundance are problematic for the nebular theory. Remember that volatile gases (which include water) couldn't condense in the inner solar system according to this model. Supposedly they were blown outward instead. Evolutionists say this is why the terrestrial planets are rocky, because lighter materials couldn't condense this close to the Sun. One of the things that shouldn't have condensed this close to the Sun is water. Obviously, that's not consistent with what we see, so where did all this water come from?

Uniformitarian scientists have long assumed that Earth's water came from the slow addition of meteorites containing hydrated minerals that came from the outer, cooler part of the developing solar system, where water was believed to be more abundant. Recent research by other secular scientists contradicts that scenario, implying that water began to accumulate on Earth right from the beginning of Earth's formation. However, in the solar nebula model, Earth started out in a hot, molten state. The high temperatures, as well as the heat from the supposed meteorite impacts would vaporize much of this primordial water, contradicting these new findings. Secular scientists have proposed different mechanisms over the years to explain the origin of Earth's water (arrival via meteorites, arrival via comets, outgassing of volcanoes, or even resulting from the collision that supposedly formed our Moon). All of these explanations have proven to be flawed. That they are still proposing new mechanisms today shows that none of these explanations are very convincing.

5.3 The Moon

The Moon is a little over 2,000 miles in diameter, about one-quarter the diameter of Earth. However, the Moon has only 1.3 percent Earth's mass. The Moon orbits Earth at a distance of about 400,000 km (250,000 miles). That is about 30 Earth diameters. It takes the Moon about a month to orbit Earth. In fact, the Moon's orbital period is the original basis for the month. We know from Genesis 1:16 that one purpose for the heavenly bodies is the marking of time. Earth's rotation period is the definition of the day, and the year is Earth's revolution period around the Sun. The day, month, and year are the natural time units, because they have an astronomical basis. The week is not a natural unit of time, but the seven-day creation account of Genesis 1 gives us the basis for the week, and Exodus 20:11 further reinforces this.

The Bible says the Moon was created for "signs and seasons" for us here on Earth and is uniquely positioned for this. As the Moon goes around the Earth, it sometimes passes directly between the Earth and the Sun. When this happens, if you're at the right spot on the Earth, you can see an eclipse. During a total solar eclipse the Moon blocks out the Sun exactly. The Moon is

Figure 5.16

Image courtesy of NASA

Figure 5.17 Apollo 11 captured this image of the Earth "rising" as seen from the Moon.

Image courtesy of NASA

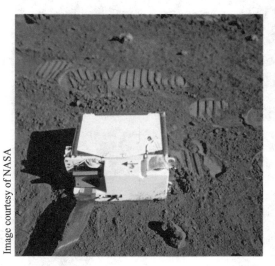

Figure 5.18 The Lunar Atmospheric Composition Experiment (LACE) was deployed during the Apollo 17 mission. Very small amounts of gases have been detected.

Image courtesy of NASA

400 times smaller than the Sun but the Sun is 400 times farther away than the Moon. This makes these eclipses possible. It also allows us to study parts of the Sun's atmosphere, things that we couldn't see otherwise thanks to the Sun's blinding light. A total eclipse is known to be one of the most awe-inspiring events in all of creation. Each time one occurs people travel from all over the world to see it. There are perhaps as many as 200 other moons in our solar system yet none of them produce eclipses like this for the planets they orbit. Only our Moon has the exact combination of size, position, and distance that produces these beautiful events.

Those who pursue knowledge of the Moon—its physical structure and composition, its landforms and geography, are engaged in a branch of astronomy known as **selenology ("Selene" is a very old name for the Moon: in Greek mythology, she was the sister of the ancient sun god Helios). **Selenography** is a sub-branch dedicated to mapping the Moon, and **selenologists** are the scientists who do all this studying. Of all the astronauts who walked on the Moon, only Harrison Schmitt, a geologist who flew on Apollo 17 in 1972 (the last manned moon mission), could be described as a selenologist as well.**

Lunar Features

The Moon's density is 3.3 g/cm³, far less than Earth's 5.5 g/cm³. Earth's crustal and upper mantle rock have density close to the Moon's 3.3 g/cm³, so the Moon cannot have a large iron/nickel core as Earth has. It is possible that the Moon may have a very tiny core, but that is not clear. For all practical purposes, we can say the Moon has no atmosphere, although technically there are trace gases that can be detected. As one might predict, the Moon also has no magnetic field.

Craters are the most common lunar surface features. *Where did the craters come from? Since astronomers discovered craters on the Moon after the invention of the telescope four centuries ago, scientists have debated that question. For much of the past four centuries, many scientists thought that volcanoes produced most lunar craters. However, others thought that craters were caused by bodies striking the lunar surfaces. In the twentieth century, most planetary scientists concluded that impacts rather than volcanoes caused the vast majority of craters on the Moon. Since the 1960s, we have discovered that many other bodies in the solar system have craters too. As on the Moon, planetary scientists think that most of the craters are the result of impacts. The impacting body is far smaller than the crater it produces. This is because the kinetic energy of the impacting body excavates most of the material to produce the crater. Kinetic energy is the energy of motion, and because impacting bodies are moving so fast, they possess huge amounts of kinetic energy. An impacting body stops when it slams into a surface, releasing its kinetic energy in an explosion that forms the crater. What were the impacting bodies? They probably were minor planets, with comets contributing a few craters.

The Moon rotates synchronously. That is, the Moon rotates and revolves at the same rate, once per month. **Synchronous rotation** causes one side of the Moon to face Earth as it orbits Earth. Before 1959, when the first spacecraft passed the backside of the Moon and sent back photographs, no one knew what the backside of the Moon looked like. Because the lunar backside was a mystery, many people called it the dark side of the Moon, meaning that it was unknown. Before Europeans had explored much of the interior of Africa in the nineteenth century, many people called Africa the Dark Continent for the same reason. Once the geography of Africa became known, the term *Dark Continent* declined. In similar fashion, the use of the phrase *dark side of the Moon* declined once we mapped the lunar far side. Unfortunately, when many people now hear the term *the dark side of the Moon*, they erroneously think that it refers to a side of the Moon that is perpetually shrouded in darkness. However, the entire lunar surface receives light from the Sun throughout the month as the Moon rotates and revolves. The only exceptions might be the bottoms of a few very deep craters near the lunar poles.*

*Looking at the Moon with the naked eye, you probably have noticed that the lunar surface has darker and lighter regions (Figure 5.20). Astronomers call the lighter regions the **highlands**, because the highlands are at higher elevation than the darker regions. The darker regions are the **maria** (ma´ ree-uh), a Latin word meaning seas. This term goes back four centuries, when astronomers began studying the lunar surface with telescopes. They thought that the maria might be bodies

Image courtesy of NASA

Figure 5.19 The far side (or dark side) of the Moon.

Figure 5.20 This image of the full Moon shows the lunar highlands (lighter regions) and maria (darker regions).

© David Rives-www.davidrives.com

of water, because the maria were dark and smooth. The word maria is plural; the singular is mare (ma´ ray). The maria and highlands have different color, because they are made of different kinds of rock. The highlands are made of rock similar to granite, while the maria consist of rock similar to basalt. Since basalt is denser than granite, the maria have sunk lower on the lunar mantle, while the highlands have risen higher, much as the continents float higher on Earth's mantle than the ocean basins do.*

*There are other differences between the lunar highlands and the maria. As previously mentioned, the maria are relatively smooth. However, the highlands are rugged, because they contain many craters. Why are there so many craters in the highlands and so few on the maria? It would seem very unlikely that objects that struck the lunar surface to form craters somehow avoided the maria. It is more likely that the entire lunar surface was struck by impacts, so what happened to the craters on the maria? A clue comes from the observation that the lunar maria appear to be circular. The isolated maria definitely are round, but the ones that overlap look like they are the intersection of several circles. When people look at the Moon through a telescope with low magnification, many of them observe that the maria look like very large craters. Astronomers think that after many of the Moon's craters had already formed several large bodies struck the Moon to produce extremely large craters. These craters were so large that astronomers call them impact basins. The impacts that formed the impact basins were so violent that they produced deep fractures that reached molten material in the Moon. The fractures acted as conduits to bring some of the molten material to the lunar surface. The molten material filled the impact basins, and in some locations spilled onto surrounding terrain. Because the volcanic material came from deep in the Moon, it has higher density than surface rocks. This caused the regions of the filled impact basins to sink to lower elevation. After the lava cooled, a few impacts on the new surface produced the few craters that we see on the maria.

The front side of the Moon is about equally divided between highlands and maria. However, the backside of the Moon is about 95 percent highlands. Why are there far more maria on the front side of the Moon than the backside of the Moon? If the impact basins formed over many millions of years as evolutionists believe, then they ought to be more uniformly distributed on the lunar surface. Evolutionists respond that the lunar crust is thicker on the backside of the Moon, preventing volcanic overflow there. However, there is a strong correspondence between impact basins and the maria, and there are relatively few impact basins on the lunar backside. However, what if the Moon is far younger than billions of years, as recent creationists believe? The impact basins would have formed in a relatively short period. Since it takes the Moon a month to orbit Earth and the maria are on one side of the Moon, we might conclude that the impacts that led to maria formation took place in a period of less than two weeks. Some creation scientists think that this might have happened at the time of the Flood.*

Evolutionists believe that 500 million years passed in between the giant impacts and the lava flows. In many places on the Moon we can see things called "ghost craters." Ghost craters are craters that are sticking up from underneath the maria. They have been partially filled in by lava but not completely erased, which is why we can still see them. These craters existed on the floors of the impact basins before the lava came and partially filled them. This means they formed *after* the giant impacts happened but *before* the lava came out from the impacts.

Here's why this is important: because evolutionists believe in billions of years they have concluded that, except for a short period early in the solar system's history, new craters form very slowly. Otherwise, there would be a lot more craters on the Moon's maria, for example. But there are many ghost craters on the Moon. Since evolutionists presuppose a slow rate of cratering, they are forced to believe that it took about 500 million years for all these ghost craters to form. However, giant impacts would have had an effect on the Moon immediately. If we look at the evidence without being influenced by evolutionary prejudices, we see evidence that cratering rates used to be a lot higher in the past than today and this affects our understanding of planets and moons all over the solar system.

Figure 5.21 The Earth and Moon to scale.

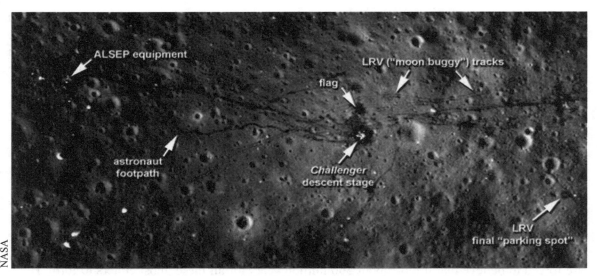

NASA

Figure 5.22 Apollo 17 mission tracks ("Challenger" was the name of the lunar lander).

Tides and the Moon

Every day in most places of the Earth, the ocean shoreline experiences **two high tides and **two low tides**. Depending on such variables as latitude, the contours of the shoreline, the depth of the water, the shape of the ocean basin, wind currents, the Moon's orbital path, and even the tilt of the Earth, the difference in the water level from high tide to low tide can vary by as little as a few inches or as much as several feet.** This is crucial to life on Earth. The Moon's tides help to circulate ocean water which prevents our oceans from becoming stagnant. This allows fish and other creatures to live there.

**Kepler had said correctly that it was principally the Moon that created the tides (Galileo scoffed at this idea, incidentally).

© Franco Volpato/Shutterstock.com

Figure 5.23 The difference between high tide and low tide can be significant in some locations.

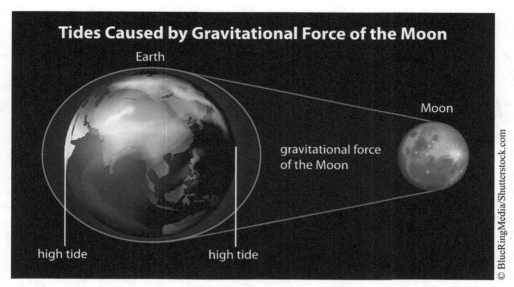

Figure 5.24 The Moon's tidal pull on Earth creates two bulges on opposite sides of the planet.

Newton showed further that the Sun, although much more massive and with lots more gravity, has an effect too, but only about half that of the Moon because it is so much further away. Newton said it was because the tidal pull of the Sun on the part of the Earth nearest it is not much more than the tidal pull of the Sun on the part of the Earth that's farthest from it. On the other hand, the Moon's pull on the near side of the Earth is quite a bit more than its pull on the far side of the Earth. This creates a bulge of water on the part of the Earth nearest the Moon, because the water has been pulled away from the Earth; and it creates a bulge of water on the part of the Earth that's farthest from the Moon too, because the far side of the Earth is closer to the Moon than the water at the **antipodes*** of the Earth, which gets left behind in the lurch, so to speak. "Antipodes" (pronounced, "an tih' poe DEEZ") refers to the opposite side of the planet, moon, or wherever you happen to be. **

When the Moon is **new or **full** (a *syzygy*), we have **spring tides**. This has nothing to do with the season of spring; they're called spring tides because the water "springs" up. Spring tides are noted by very high high tides and very low low tides. In this situation the Moon's and Sun's gravitational fields are working together to create two very large tidal bulges on opposite sides of the Earth. **We have spring tides twice a month.** When the Moon is at *quadrature* (1st **quarter** or 3rd **quarter**), we have **neap tides:** not very high high tides and not very low low tides. The Moon's and Sun's gravitational fields are working at right angles to each other. **We also have neap tides twice a month.****

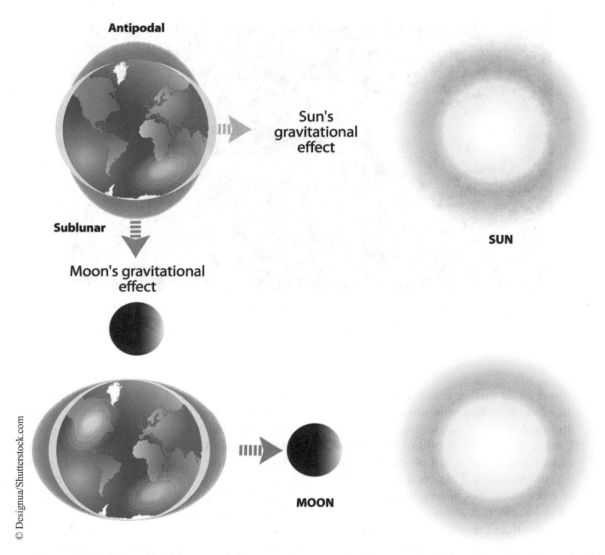

Figure 5.25 When the Moon and Sun are at *syzygy*, we have spring tides. When the Moon is at quadrature, we have neap tides.

Lunar Origin Stories

We see in Genesis that the Moon was made to mark out times and seasons and to rule the night. However, naturalism rules out any discussion of creation on philosophical grounds, so evolutionists were very excited about the Apollo Moon program back in the 1960s. They thought that the lunar missions would help them understand how the Moon could have formed by itself without a Creator being involved.

Apollo 17 astronaut Harrison Schmitt was not only the last astronaut to walk on the Moon, but he was also a trained geologist. Schmitt and the other astronauts gathered samples from the Moon's surface and brought back hundreds of pounds of Moon rocks with them. These rocks revealed some very interesting things about the Moon. At the time there were three main evolutionary theories for the origin of the Moon—the fission theory, the nebula theory, and the capture theory. Evolutionary scientists were eager to analyze the Moon rocks and see which of their theories was right.

The fission theory said that the early Earth was spinning so rapidly that a chunk of material tore off and became the Moon. However, tests showed that lunar rocks are different than Earth rocks in some important ways. For example, the Moon is deficient in iron when compared to the Earth. There are also serious problems with the physics of such a rapidly spinning Earth.

The nebular theory said that the Earth and Moon formed out of the same swirling cloud of gas and dust. This meant the Earth and Moon were formed from the same materials, but the Apollo missions showed that Moon rocks are different than Earth rocks. There are other problems built into the solar nebula model as mentioned earlier.

The third theory was the capture theory, which said that the Moon formed elsewhere and one day it passed too close to the Earth.

Image courtesy of NASA

Figure 5.26 Collecting lunar samples.

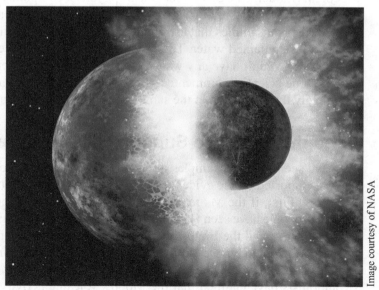

Image courtesy of NASA

Figure 5.27 An artist's rendering of the large impact.

The Earth's gravity pulled it in and the Moon started orbiting our planet. There are several problems, the biggest one being that one object can't capture another one gravitationally like this without interacting with other objects. If it were just the Moon approaching the Earth then Earth's gravity would have accelerated the Moon on the way in. By the time the Moon arrived it would be going too fast to stay in orbit, so either the Moon would hit the Earth or else it would zoom on by and get flung away. Either way you don't get an orbit.

So the Apollo missions left evolutionists in turmoil. Each theory had fatal flaws. A few years after the Apollo missions ended a NASA publication said this: "in spite of everything that we have learned during the last few years we still cannot decide between these three theories. We still need more data, and perhaps some new theories, before the origin of the Moon is settled to everyone's satisfaction."

A new idea—the "large impact hypothesis"—soon came to the rescue, claiming that billions of years ago, an asteroid the size of Mars crashed into the Earth. Both the asteroid and the Earth were broken up in this catastrophe. Lots of material was sprayed out into space. Some of it formed into the Moon and the rest of it came back down into the Earth. However, the iron from both bodies sank into the Earth's core so this explains why the Moon doesn't have much iron today. Theorists have named this object that supposedly struck the Earth "Theia," as if naming the unseen object of their speculation lends greater credibility.

Is this theory any better than the others? Some claim that computer simulations have proven it to be true, but computer simulations can never prove anything happened, only that it's one possibility for how something might have happened, given the assumptions built into the program. Secondly, this model only works within very narrow parameters. You need an asteroid of exactly the right size coming in at exactly the right speed and exactly the right angle, and there is debate about whether or not it could work even within these narrow boundaries. Even many evolutionists don't accept this model. One evolutionary astrophysicist complained, "the collision has to be implausibly gentle" to avoid messing up Earth's orbit. Not only that, scientists recently took a closer look at some of the soil samples that astronauts brought back from the Moon, and discovered that some of the samples contained water. This was a complete shock to evolutionists. The impact theory says that the Moon can't have any water today. It would have been vaporized and lost during the collision. As one scientist commented, "it's hard to imagine a scenario in which a giant impact melts completely the Moon and at the same time allows it to hold onto its water."

Does the Evidence Support Billions of Years?

We have seen several examples already of how the predictions based on billions of years are not always supported by the evidence. As we close our discussion about the Moon, we should consider two more. First, if the Moon is billions of years old, we expect it to be geologically inactive. For centuries people have seen flashes and temporary glows of light in lunar locations such as the Aristarchus plateau and the Eratosthenes crater. These events don't last long so it's been very difficult to photograph them but we have written accounts of them going back to the 1500s. It seems that the Moon is still geologically active, releasing gas from volcanic vents and experiencing moonquakes. Evolutionists have claimed that the Moon's volcanic activity stopped about 3.2 billion years ago, and that it can't be geologically active. Eventually the accumulating evidence became too much, and in 1968 NASA released its catalog of reported lunar events which listed over 500 separate sightings of anomalous events on the Moon. A couple of years later the Apollo 15 mission measured radon gas coming from the Aristarchus crater. The Moon still shows signs of geologic activity, which makes sense if it is only thousands of years old.

There is one other problem worth mentioning. When astronauts went to the Moon, they brought experiments with them. One of those experiments was called LLR or lunar laser ranging. These were special mirrors that the astronauts of Apollo 11, 14, and 15 left behind on the Moon. Scientists on the Earth can fire a laser at one of these reflectors and if the laser hits the target directly, then the laser bounces back to Earth. Scientists then measure how long the light took to return and then they figure out exactly how far away the Moon is from us. This is still being done today.

Scientists have discovered that the Moon is moving further away from us every year. This was already suspected before Apollo but now we've been able to measure it. This is happening because the Moon is raising ocean tides on the Earth. This causes our ocean to bulge toward the Moon. However, the Earth is spinning underneath the bulge and pulling it forward. The bulge itself then exerts a gravitational pull on the Moon which accelerates it and moves it outward in its orbit. That's a complicated process, but the important part is this: the Moon is moving away from the Earth at about 3.8 cm (about 1.5 in) per year. We've measured this for years and we know why it's happening. An inch and a half per year doesn't sound like much but the way this works, it would have been receding faster in the past. Looking backward in time the Moon would have been touching the Earth only one and a half billion years ago, which of course is impossible.

5.4 Mercury

Mercury is the innermost planet of our solar system, as well as the smallest, with a diameter about one-third that of the Earth. Mercury was the messenger of the gods in Roman mythology—the guy with the winged feet. The name fits, as per Kepler's third law of planetary motion, Mercury orbits the Sun faster than any planet, taking just eighty-eight days for one revolution. As speedy as Mercury may be revolving around the Sun, it completes one rotation in a little over fifty-eight and one-half Earth days. However, that would be a sidereal day. Because the planet is also revolving so rapidly a solar day is 176 days long! That means that a given spot on Mercury is in direct sunlight for about eighty-eight continuous Earth days at a time. It may seem peculiar but the same planet that has the shortest year also has the longest solar day.

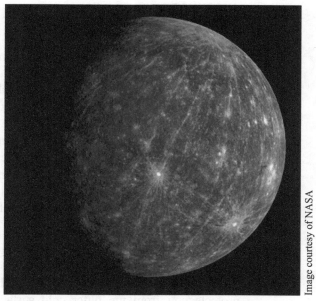

Image courtesy of NASA

Figure 5.28 This view of Mercury was captured by the *MESSENGER* spacecraft.

Image courtesy of NASA

Figure 5.29 Multiple views of Mercury, using infrared imaging from *MESSENGER*.

Being roughly 0.4 AU from the Sun, Mercury has always been seen either leading the way to a sunrise or chasing after a sunset, but never straying far above the horizon. Mercury does have the most eccentric orbit of any planet—meaning that its orbit is noticeably elliptical and not as circular as the other planets. However, due to the Sun's glare, it is a challenge to see Mercury even when it is above the horizon, and it is difficult to study from the Earth. Mercury is only visible at twilight and only at certain times of the year when it is in a part of its orbit that appears (in angle) most distant from the Sun. This position is called "greatest elongation." At such times, it is possible to see Mercury just after sunset for "eastern elongations" or just before sunrise for "western elongations." There is a long history of frustrated astronomers who wanted a much better look at Mercury, but that would require a view from space rather than Earth.

Mercury has no seasons, as unlike the Earth its axis is not tilted. It is however a planet of extremes due to (1) its slow rotation and (2) its lack of atmosphere. Technically, there is a very thin atmosphere, but for all practical purposes we can treat Mercury as if it has none. During the daylight hours, the surface temperature reaches nearly 800°F, while the nighttime temperatures plunge to nearly negative 300°F. These are the greatest extremes of any planet. The incredible daytime heat may come as no surprise, but the cold nights may. Without a sufficient atmosphere, there is no "greenhouse effect"— no trapping of the daytime heat, so it radiates away at night. Anyone who claims that we need to "stop the greenhouse effect" here on Earth has no idea what they are talking about!

Creation astronomer, Dr. Jason Lisle makes this interesting observation regarding Mercury's temperature extremes:

> "These extremes lead to some interesting hypothetical scenarios. Suppose your spaceship ran out of fuel, but not before you were able to land near the equator of Mercury. Thankfully, you have landed on the night side (barely), and your spacesuit protects you from the bitter cold surface. But it won't protect you from the 800-degree daytime temperatures that are about to occur when the sun rises in just a couple of hours! Thanks to Mercury's very long day, you could begin jogging

west at a leisurely pace of two miles per hour. As long as you can maintain that pace, you could stay ahead of the sunrise and remain safely in the night for as long as it takes for the rescue ship to arrive. Effectively, you would be jogging in the opposite direction that the planet rotates, and at about the same rate, thereby permanently remaining in the safe shadow of night."

Closer Looks and Unexpected Surprises: Mariner and Messenger

We got our first good images of the surface of Mercury when the space probe *Mariner 10* made two passes in 1974 and a third in 1975. Mercury looked a little bit like Earth's Moon, with lots of craters, mountains, and signs of lava flows. Mercury also had a network of cliffs called lobate scarps. The most impressive feature was the 900-mile wide impact basin named *Caloris Basin* ("Basin of Heat").

The *MESSENGER* spacecraft made several passes of the planet and then assumed an orbit around the planet for a four-year mission to do a thorough, close-up study of the planet. The mission continued until 2015, when *MESSENGER* ran out of fuel and was purposely crashed into the planet. As is often the case, these leaps forward in the ability to make observations raised more questions than they answered.

Mariner 10 had shown that Mercury was so dense that most scientists believe it has an iron core occupying over 40 percent of its volume. This poses a huge problem for secular theories. Evolutionary models say Mercury can't be this dense. As one secular astronomer explained, "the driving force behind previous attempts to account for Mercury has been to fit the high density of the planet into some preferred overall solar system scheme." Today this problem is "solved" by proposing

another large impact hypothesis (as we saw with the Moon). Evolutionists believe that Mercury did actually form in line with the solar nebula model. However, after the planet formed, a large asteroid crashed into Mercury and stripped away all the lighter material from it. The lighter stuff was ejected into space, leaving behind the dense material in the planet today.

The *Mariner* 10 probe also discovered that this little planet has a magnetic field but according to evolution it can't have a magnetic field. To understand why this is important we need to discuss magnetic fields for a moment. There are several reasons why a planet could have a magnetic field but most of them require the planet to be young. Since evolutionists

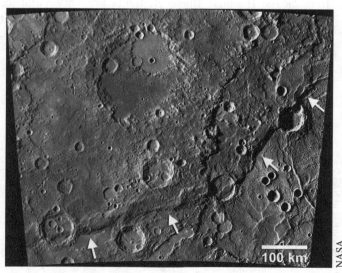

Figure 5.30 Enterprise Rupes, largest lobate scarp on Mercury (white arrows).

NASA

Figure 5.31 False-color image of the Caloris basin.

believe that the planets are all billions of years old, this means there can be only one source of magnetism for the planets: a dynamo deep inside each planet. A dynamo means there is hot liquid metal moving around inside each planet. As it flows an electrical current is produced which creates a magnetic field. However, in order for a planet to still have a magnetic field after billions of years it has to have a liquid core, but Mercury is so small that the general secular opinion is that the planet should have frozen millions, if not billions of years ago. What we observe about Mercury does not fit the evolutionary model.

When the MESSENGER spacecraft arrived at Mercury, we got a treasure trove of new information, and several new challenges for secular models. First of all, MESSENGER found that Mercury has a lot of volatile elements. This discredits the large impact hypothesis. Because these elements are very volatile, a violent collision like this would have vaporized these elements. They would have escaped into space and they wouldn't be on Mercury today. The presence of volatiles refutes the notion of a large impact, and without a large impact there is no secular explanation for the density of the planet.

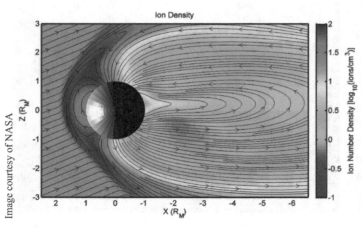

Image courtesy of NASA

Figure 5.32 Mapping out Mercury's magnetic field.

Second, Mercury appears to have lots of sulfur, a higher concentration than any other rocky planet. This sulfur causes other problems for evolutionary models. The nebula theory says that volatile elements like sulfur can't be there (even before an impact). As one study noted chemical condensation models indicate that sulfur cannot condense in the primordial solar nebula at the heliocentric distance of Mercury.

Third, MESSENGER found that Mercury's magnetic field is decaying. (Actually, the decaying field was not a surprise to creationists because a physicist named Dr. Russell Humphries had predicted it ahead of time based on the creation account in the Bible.) At its current rate of decay it loses half its strength about every 320 years. If Mercury were really billions of years old it shouldn't have a magnetic field at all, and now we see that it not only has a field, but the field is also decaying rapidly. Those are the marks of a "young" planet. It can't have been decaying like this for billions of years because it would have been impossibly strong far less than 1 million years ago.

Lastly, MESSENGER found other evidence that supports the idea that Mercury is young. One surprise was a completely new feature on Mercury. The "blue hollows" are depressions up to a few miles across. Scientists think these hollows form when volatiles escape Mercury's surface in various ways. Many of these hollows appear quite fresh, so apparently, whatever process creates them is still going on today. Again if Mercury were really billions of years old, these processes would have ceased long ago. These hollows make Mercury look quite young.

To sum up the ongoing problems, the solar nebula theory couldn't have predicted how dense Mercury really is. The solar nebula theory would not have predicted a magnetic field. The solar nebula model could not predict the levels of volatile elements we find on Mercury. The large impact hypothesis cannot save the day! The rate of decay of the magnetic field and the observations of the blue hollows point to an object that is much younger than the nebular model could predict. The observations do not match the evolutionary model at this time, but they do make sense in a creation model.

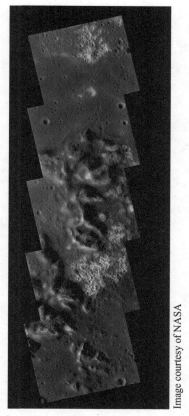

Image courtesy of NASA

Figure 5.33 Mysterious hollows on Mercury.

The Solar System— Part 2

Chapter 6

> *O LORD, our Lord, How majestic is Your name in all the earth, Who have displayed Your splendor above the heavens!*
>
> *When I consider Your heavens, the work of Your fingers, the moon and the stars, which You have ordained;*
>
> *What is man that You take thought of him, And the son of man that You care for him?*

—from the 8th Psalm

Figure 6.1 Venus.

Figure 6.2 Mars.

Figure 6.3 Jupiter.

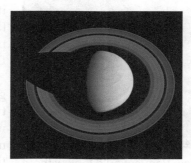

Figure 6.4 Satum.

Images courtesy of NASA

6.1 Venus: Earth's Twin

If you consider bright objects in the night sky to be beautiful then it comes as no surprise that Venus was named for the Roman goddess of beauty. Pagan cultures worshipped the bright wanderer. Blake, Frost, and Wordsworth were inspired to write poems about "her," and van Gogh included Venus as a symbol of loveliness in three of his paintings. After the Sun and Moon, no other object is as bright as Venus as viewed from Earth. In part Venus dazzles us because it is the planet closest to Earth, but more importantly Venus is blanketed with highly reflective clouds. This ability to reflect light is called "albedo," and the value of 0.75 for Venus is the highest of any planet. What the 0.75 tells us is that Venus reflects 75 percent of all the Sunlight that hits it, compared (for example) to our Moon at about 10 percent or an albedo of 0.1.

Like Mercury, Venus is either seen ahead of the rising Sun or chasing after the setting Sun, giving it the titles of the "morning star" and the "evening star." At a distance of just over 0.7 AU from the Sun, Venus is able to distance itself from the glare, but it is still held in a wedge on either side of the Sun. At "greatest elongation," Venus will appear to be at its furthest from the Sun (about 45 degrees), giving naked-eye observers several hours to enjoy the sight before the Sun rises or after the Sun sets. Venus is bright enough that we can still see it within about 5 degrees of the Sun. Recall from Chapter 3 that Galileo discovered that Venus actually goes through phases, which was support for heliocentrism. In a telescope, as Venus gets closer it looks bigger and brighter, even though it appears as a tall crescent. As Venus moves further away it approaches its "full" phase but it looks much smaller and is actually less bright as viewed from Earth.

A cursory comparison of Venus and the Earth gives rise to the nickname of "Earth's Twin." The two planets are nearly the same diameter and mass, with Earth being just a bit larger. Both planets have nearly circular orbits, and of course they enjoy relatively close proximity by astronomical standards. They also both have atmospheres, something poor

Figure 6.5 Venus shrouded in its clouds.

© Dotted Yeti/Shutterstock.com

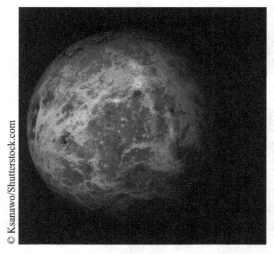

Figure 6.6 How Venus would look without its thick cloud cover.

© Ksanawo/Shutterstock.com

Mercury was lacking for the most part. However, a closer look at Venus suggests that if it is "Earth's Twin," then in some ways it may be thought of as Earth's evil twin.

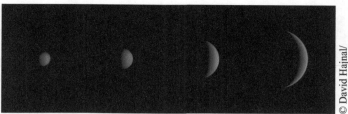

Figure 6.7 Venus looks largest and brightest in its crescent phase.

The Russians actually landed a number of spacecraft on the surface of Venus in the 1970s and 80s, but their time for making observations was often just minutes to an hour. Venus is the hottest planet in the solar system, reaching temperatures of close to 900°F, high enough to melt some metals. However, unlike Mercury there is no cooling down at night. Even though Venus is not as close to the Sun as Mercury, its incredibly dense atmosphere creates a pressure over ninety times that of Earth's atmosphere. That's roughly the pressure you would experience nearly a mile under the ocean. It is crushing. The atmosphere is over 95 percent carbon dioxide—Earth has only trace amounts, so what we are observing is the greenhouse effect on steroids. The term typically used by scientists is "the runaway greenhouse effect." If the hellish temperatures and crushing pressure weren't enough, there are clouds that rain sulfuric acid. The good news is that this acid rain evaporates in the heat before it reaches the planet's surface.

Calling Venus our "evil" twin may sound unkind, so perhaps "quirky" is better. One need look no further than the rotation of the planet to see more quirkiness. Venus rotates backward. The Sun would appear to rise in the west and set in the east. This rotation is painfully slow, with one rotation (sidereal day) taking 243 Earth days, which is longer than the 225 Earth day period of revolution. In addition, because of the backward rotation, the sidereal day is longer than a solar day (117 Earth days) which is the opposite of what we saw with the Earth and Mercury. So while Venus has the least eccentric orbit of any planet (nearly circular), it has very unusual rotation characteristics.

Getting a glimpse of the surface of Venus from Earth has proved quite challenging because Venus is perpetually blanketed in such thick clouds, concealing all surface features. In 1990, the NASA spacecraft *Magellan* was placed into orbit around Venus and began mapping the surface using radar. Since radar is not blocked by clouds, *Magellan* was able to map the topography of Venus in unprecedented detail. The four-year mission revealed that Venus has many of the same geological features found on Earth, including mountains, valleys, canyons, volcanoes, lava flows, and plains. *Although there are no oceans of liquid water, Venus does have two "continents," highland regions made of lighter granitic material: *Aphrodite Terra* and *Ishtar Terra*. Neither continent is particularly high in elevation, although Maxwell Mountain in Ishtar is a couple of miles higher than Earth's Mt. Everest. Continental land masses make up only about 10 percent of Venus' topography; 20 percent is low flatlands and 65 percent is rolling upland plains.*

Thanks to *Magellan*, we also learned that Venus has some of its own, unique geological features as well, such as arachnoids, tesserae, coronae, and pancake domes. The Soviets first

Source: NASA

Figure 6.8 Computer-generated image of 5-mile-high Maat Mons.

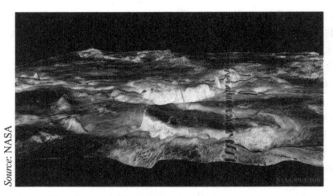

Source: NASA

Figure 6.9 Cylindrical coronae on Venus.

Source: NASA

Figure 6.10 108-mile-wide Isabella, second-largest impact crater on Venus.

named the arachnoids–regions where the terrain has been folded and broken into a gossamer-looking structure resembling a spider web. They are circular structures, ranging from 50 km to 230 km (30 miles to 138 miles) in diameter with a central volcanic feature surrounded by a complex network (or "web") of fractures. Tesserae are complex ridged terrains found on plateaus. "Corona" is from the Latin meaning "crown." They are oval-shaped features thought to be produced by plumes of rising magma currents that cause the surface to bulge and then collapse in the center, forming ornate, concentric rings. Pancake domes are similar in most respects to shield volcanoes on Earth, such as those found on the Hawaiian Islands. However, pancake domes are flatter and broader than their terrestrial counterparts.

*There are also **impact craters** on Venus, but they are mostly of the mid-size range (averaging about 25 to 35 miles across). That's because Venus' thick atmosphere destroys any potential small impactor that might leave a small crater, while large impactors are broken up by the collision with the atmosphere and end up making medium-sized craters instead of big ones.* However, the largest crater on Venus is still 170 miles across.

*The rocks on the Venusian surface, as seen in images taken by the Russian *Venera* spacecraft, appear similar to earthly basalts—dark volcanic rocks common in Earth's oceanic crust. But that's where the similarity ends. Venus is a desert wasteland, with thick, heavy, scorched unbreathable air; the sky is orange in color; it's been suggested (although not universally accepted) that there are also powerful lightning storms on the surface.* Earth's twin also has no moon and virtually no magnetic field.

Models, Predictions and Observations

According to the solar nebula model, Venus was formed by the same processes as the Earth, at the same time, in about the same

Figure 6.11 Venera view of the surface of Venus (landing pad at right bottom).

place, and from the same materials. Therefore, Venus *should* be very similar to Earth but as we've already seen, Venus is actually very different. However, many astronomers argue that the two planets were much more alike in the past. For example, some planetary astronomers suggest that Venus *did* have a Moon initially, but then something happened. Yes, that's right—another "large impact." An asteroid came along and hit it and destroyed Venus's Moon. The lack of evidence for this claim has many astronomers rightfully silent on whether or not Venus ever had a moon.

Why does Venus rotate backward? One popular explanation has been yet another "large impact theory." However, many astronomers are giving up on that particular explanation and exploring mathematical models to suggest that somehow the gravitational influence of the Sun may have resulted in the retrograde rotation.

Venus's surface appears to be quite young. The entire surface appears to be fresh. There's no record anywhere of billions of years of erosion or chemical weathering. Evolutionists are forced to speculate that somehow, for reasons unknown, the entire planet was resurfaced by volcanic activity not that long ago. This means the entire planet's crust would be submerged under lava all at the same time. Most planetary scientists believe Venus was resurfaced "only" about 500 million years ago. They believe this based on the number of impact craters on its surface and the assumptions built into their model. They're making two big assumptions here. First, they're assuming they know how often new craters form today and second, they're assuming this rate has been the same for at least 500 million years. However, both of these assumptions have been shown to be invalid. Studies on our Moon and elsewhere in the solar system have shown that the evolutionary cratering model is fatally flawed.

Using the solar nebula model, astronomers made a number of predictions about Venus prior to our ability to launch space probes to check those predictions. Based on the model, Venus should have had the same internal geological structure as the Earth and thus a similar magnetic field, as well as plate tectonics. For all practical purposes, there is no magnetic field and there is no evidence of plate tectonics despite the widespread volcanic features. The observations do not support the predictions of the solar nebula model. One popular, secular, astronomy college textbook summed it up this way:

> "Venus seems to have followed a peculiar path through planetary development, and its history is difficult to understand. Planetary scientists are not sure of all the details about how the planet formed and differentiated, how it was cratered and flooded, or how its surface has continued to evolve."

Could it be that the mysteries, peculiarities, and contradictions exist because of an a priori commitment to naturalism and billions of years? Perhaps Venus looks "young" because it is?

6.2 Mars—The Red Planet

At 1.5 AU from the Sun, Mars, the "red planet," has been a cultural obsession since it was worshipped as a god millennia ago. Named for the Roman god of war, the planet more recently gave inspiration to science fiction writers of the 1800s and early 1900s. The obsession continued to grow in the form of books, films, and television programs, most of which pointed to Mars as a bastion of intelligent—and perhaps dangerous—life. If science fiction fans are preoccupied with Mars, it is no less the case with NASA. A steady stream of funding has followed the promise of finding life on Mars. The idea of intelligent life has long been abandoned, and even the notion of finding simpler forms of life today is met with skepticism, so now what NASA hopes to deliver is proof that life once existed on Mars in the past.

*The year 1887 was a banner year for Mars, as our Earth passed the red planet during a **perihelic opposition**, a little under 35 million miles away. A few years before, a large 26-in refractor was installed at the U.S. Naval Observatory on the outskirts of Washington, DC (still in use today!). With it, Observatory director **Asaph Hall** discovered **Phobos** and **Deimos**, the two small moons of Mars.*

Figure 6.12 The red planet.

Figure 6.13 U.S. Naval Observatory's 26-in refractor.

*Meanwhile the Italian astronomer **Giovanni Schiaparelli**, director of the Milan Observatory, observed what he called "**canali**" on the Martian surface. "Canali" is the Italian word for "**channel**," and all he meant by it was that he saw some straight lines on Mars that might be natural features. But in America, folks thought he meant, "**canals**," which of course are not natural at all, but man-made (or in the case of Mars, *creature-made*.) The idea that there might be intelligent beings on Mars electrified the world.

> "... the broad physical conditions of the planet are not antagonistic to some form of life; ... there turns out to be a network of markings covering the disk ... this ... may be [coincidence], ... but the probability points the other way."

—**Percival Lowell**

Percival Lowell was a very rich Bostonian businessman, and one of his passions was astronomy. He built an observatory in Flagstaff, Arizona, and observed the red planet a few years after the close passing of Mars. Some astronomers besides Lowell also claimed to see canals, however most of the scientific community dismissed his work as the result of an overactive imagination. But Lowell was rich. Just because he couldn't get his work published in scientific journals was no real impediment. He simply published his own books, beginning with a work titled *Mars* in 1895, and he went on the lecture circuit, speaking directly to the public.* After his death, the Lowell Observatory continued its work, which included the discovery of Pluto.

*Lowell speculated that the canals were a great engineering work designed to bring water from the polar caps of Mars to the warmer equatorial regions in order to irrigate the crops of a cold, dry planet—a last ditch effort by Martians to stay alive on a dying world. A Mars mania swept the globe, and was even the inspiration for the H.G. Wells classic, "The War of the Worlds," as well as Edgar Rice Burroughs' John Carter of Mars stories.

The Mars mania created by Lowell and others led to a real Mars scare when on Halloween night in 1938, H.G. Wells' classic story, "The War of the Worlds," was turned into a radio broadcast that featured "news-bulletins" detailing the developments of an invasion of New Jersey by the Martians. Newspapers alleged that thousands panicked, fleeing the cities, contemplating suicide, and even volunteering to fight the aliens,* and the career of future Oscar-winner Orson Wells nearly ended prematurely for his involvement with the broadcast.

© Everett Historical/Shutterstock.com

Figure 6.14 Percival Lowell.

Figure 6.15 Scale size comparison.

Mars is about one-half the diameter of Earth, and about one tenth the mass. Its 24½-hour period of rotation is just slightly longer than ours, and at 24 degrees, it has a similar axial tilt, which means that Mars also has seasons like Earth. Its year is nearly twice as long as an Earth year at 687 Earth days however, so each Martian season lasts roughly half an Earth year. Temperatures vary from a brisk −200°F to a balmy 60°F, with the average temperature at just −80°F.

Mars has polar ice caps that are visible from Earth using a small telescope. These ice caps become more expansive in the winter and shrink in the summer in their respective hemispheres—just as we see with the ice caps on Earth. While Earth's ice caps are water-ice, Mars' ice caps are water-ice layered underneath several feet of frozen carbon dioxide (dry ice). Carbon dioxide also makes up 95 percent of the Martian atmosphere, but that atmosphere is just 1 percent as abundant as Earth's atmosphere.

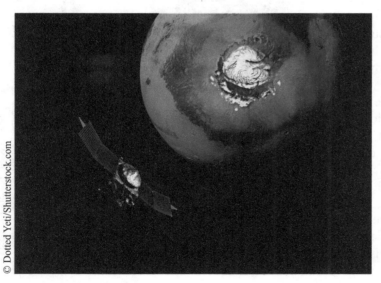

Figure 6.16 Illustration of South polar ice cap of Mars, with MAVEN spacecraft.*

*On tiny Mars there are immense mountains and great canyons. **Olympus Mons (Mount Olympus)** is a dormant volcano roughly fifteen miles high, roughly three times the size of Mount Everest, Earth's largest mountain.* Olympus Mons is the largest volcano in the solar system. It's about the same size as the state of Arizona. *Girdling the Martian equator, a great crack in the ground called the **Valles Marineris (the Mariner Valley),** stretches across thousands of miles, about the distance between the east and west coasts of the United States.* This enormous rift valley—the largest canyon in the solar system—is about ten times as long as the Grand Canyon, about seven times as wide, and four times as deep. It's almost as deep as Mount Everest is high. Astronomers believe that the planet's structure is differentiated somewhat like the Earth, with a core, mantle and crust. It is believed that Mars has a very thick crust, which allows it to support such massive shield volcanoes as Olympus Mons.

The Martian surface is a mixture of yellow, red, and orange, the result of rusting in the metal-rich soil. The generally red soil is stirred up by seasonal dust storms, giving Mars an even more pronounced red appearance and explaining why we call it the Red Planet. One particular metal, **hematite**, which was picked up by *Opportunity* on the Meridiani planum, is an oxide of iron with the same chemical makeup as rust, Fe_2O_3. The red color that results from these minerals and wind storms is easily discernible from Earth, even without a telescope. *The rovers— *Sojourner, Spirit, Opportunity, Curiosity*—have trekked miles across the boulder-strewn plains, climbed 1,500 foot hills, and descended (and climbed back out of) some fair-sized craters. Besides the usual collection of dust and loose rocks on the surface of Mars, solid bedrock has been found.*

The Two Moons of Mars— Captured Asteroids?

*In orbit above the red planet are two small, cratered moons, **Phobos** and **Deimos**.* Phobos is the larger of the two—about 10 miles in diameter. Deimos has a diameter of only 8 miles. *Because of their small size and negligible mass, they are oddly shaped. The Sun, planets and larger moons of the solar system are all round because their gravities squeeze them into the most energy-efficient shape—a sphere. But if an object is small enough and light enough, it can be pretty much any shape at all. There are hundreds more small, irregular moons in the outer

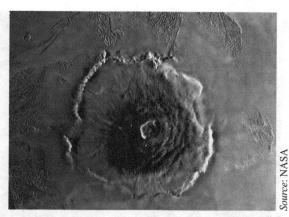

Source: NASA

Figure 6.17 Olympus Mons (Mount Olympus).

Source: NASA

Figure 6.18 Valles Marineris, the Mariner Valley.

Source: NASA

Figure 6.19 Viking 1 landing site.

Figure 6.20 Mars and its two moons.

solar system—at Jupiter, at Saturn and beyond. And there are thousands of objects like these in the asteroid belt.*

Phobos ("Fear") orbits closer to its planet than any other moon at an unbelievably close distance of only 3,700 miles above the surface. Its proximity to Mars—combined with Mars' gravity—means that Phobos orbits very quickly, completing one orbit in seven and one-half hours. A greater distance away from the surface, Deimos ("Panic") takes just over thirty hours to complete one orbit.

From a secular perspective, the origin of these moons is perplexing. It may be that they were once asteroids that at some point were captured by the gravity of Mars. This is possible, but it involves an improbable chain of events. Moreover, captured asteroids are expected to have exaggerated, elliptical orbits, but Mars' moons orbit in nearly perfect circles.

Figure 6.21 Phobos.

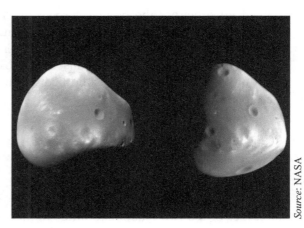

Figure 6.22 Deimos.

Life on Mars? Past, Present, and Future

*More spacecraft have traveled to Mars than to any other planet. Of the dozens that went, several have succeeded in reaching Mars intact and functioning. There have been orbiters and landers and rovers—and these *Mariners, Vikings, Pathfinders, Surveyors,* and their successors have sent back amazing images and volumes of information. One of the jobs given to these spacecraft was to find out if life might exist on Mars.* If you were paying attention in 2021 when *Perseverance* landed, then you know that the purpose of this mission is to find any signs of life. Not intelligent life, not current, simple life forms, but any remains, remnants or other indicators that there may have once been life on Mars in the past. Despite what the folks at

Figure 6.23 *Perseverance* successfully landed in 2021.

Source: NASA/JPL-Caltech

NASA might contend, if their mission is a "success," (1) it does not prove evolution is true, (2) it does not disprove biblical creation, and (3) it should face well-deserved scrutiny.

What you may not know is that NASA has already made claims about life on Mars. In 1996, NASA put out a report titled "METEORITE YIELDS EVIDENCE OF PRIMITIVE LIFE ON EARLY MARS." Their team of researchers was adamant that they found evidence of life in the form of microscopic fossils. The problem in the end was that no one outside of NASA agreed. Other secular scientists debunked the claims. In 2007, the respected scientific journal *Nature* ran an article with this headline: "Wheel of *Spirit* Hints at Life on Mars." *Spirit* was one of several rovers that we landed on Mars, and if all you saw was this headline you'd think *Spirit* found evidence for life on Mars. What *Spirit* actually found was a small patch of sandy material which might be silica, which might have formed from a now extinct volcanic fumarole. On Earth volcanic fumaroles sometimes have thermophilic bacteria in them, so astrobiologists try to stretch this into a claim that we found evidence for life on Mars. Of course, telling people that a patch of sand hints at life on Mars is misleading at best. However, secular researchers do this sort of thing pretty often, coming up with wild speculations which are often presented as scientific facts. If/when another announcement is made about life on Mars, you have good reason to be skeptical, and you have no reason to think that the discovery of life—real or imagined—changes the status of either creation or evolution. Evolution can't explain how life could have started without a Creator, but evolutionists are always looking for life somewhere else as if that would help them solve this problem. However, finding life somewhere else would just increase the number of places where evolution can't explain its origin.

Not to belabor the point, but NASA—an agency that relies on government funding—often operates in a world where their current missions and future projects have to be "hyped" with exaggeration, speculation, and even denial of unwanted realities. For example, on the subject of extra-terrestrial life, NASA boldly claims that "the universe is teeming with life," despite the absence of any supporting evidence. Meanwhile, reputable astronomers at places like Harvard University (which does not rely on government funding) acknowledge that *if* there is life out there, we may never find it. Then, in what may be viewed with some humor, as recently as 2019 the NASA chief has insisted that "Pluto is a planet." NASA launched a mission to Pluto early in 2006, only to see Pluto demoted

Figure 6.24

later that same year. For whatever reason, NASA seemed to take this very personally, so even though the rest of the astronomical world has moved on, NASA clearly has not! One final example would be the manned mission to Mars. While earlier releases from NASA and media coverage might have people thinking this will happen soon, that simply is not the case.

Mars has no magnetic field. There is speculation as to how or why that is the case (including impact theories), but with all the discussions about sending astronauts to Mars, this is one of several glaring problems for such a mission. The trip itself is expected to take seven months—that is seven months of exposure to harmful radiation in space. After reaching Mars, aside from its cold, dry environment, lack of oxygen, and huge dust storms, there's still the matter of the harmful radiation. Between the loss of its magnetic field and its thin atmosphere, the surface of Mars is exposed to much higher levels of radiation than Earth, and those levels could be lethal. Anyone participating in the mission risks acute radiation sickness, increased risk of cancer, genetic damage, and even death. While many good scientists and engineers are trying to figure out how to address the problem, it is a huge, costly obstacle.

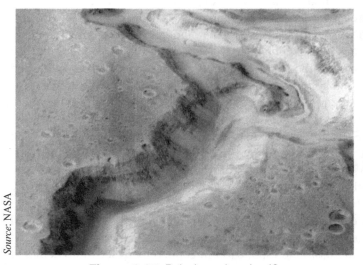

Source: NASA

Figure 6.25 Dried-up river bed?

Water, Water . . . Everywhere?

Despite the hopes and expectations of Percival Lowell, Edgar Rice Burroughs, and science fiction fans everywhere, life has never been found on Mars. But orbiting probes and robot rovers have uncovered evidence that liquid water may still be found deep underground, and other pictures reveal what appear to be dried-up river valleys, flood zones, alluvial fans, and sea or lakebeds—signs, they say of a more Earth-like Mars in the distant past. Perhaps Mars once had life, long ago, and perhaps even now simple forms of life might exist on, or under, the dry, frozen wastelands of Mars. Claims about the importance of locating water are not limited to Mars. The logic is that if you can find water you can find life. Of course there is no logical necessity for such a claim. Life as we know it may rely on water, but water can exist without life.

Some evolutionists say large parts of Mars used to be flooded with water, maybe even the whole planet. How could there have been flooding on Mars? Mars does actually have water on it today but not in liquid form. For example, there's water vapor in the thin atmosphere. More importantly, Mars has some water ice at the poles. Water is also fro-

Figure 6.26 Dried-up lake bed?

zen under the surface elsewhere in the planet. This would seem to be the source of at least some of the formations we've seen. For example, a volcano or some other event that heats up the surface sufficiently could melt the ice frozen into the surface and cause flooding but how long could this liquid water last on the surface under current conditions? It can't last long at all, which is why Mars is so dry today. Because Mars has a very thin atmosphere, the boiling point of water is very low. Even though it's very cold on Mars a sample of liquid water would evaporate away quickly. Despite all this, evolutionists speculate that Mars used to have huge oceans of liquid water which lasted for hundreds of millions of years.

Evolutionists believe that Mars used to have a thick atmosphere but then the combination of the solar wind and the loss of its protective magnetic field resulted in slowly stripping away the atmosphere, and that resulted in the eventual loss of its oceans of water. Some theorists have proposed large impact scenarios to explain the loss of atmosphere as well. Regardless, what we observe today is that Mars is a desert planet where it's physically impossible to have liquid water, and the claims about oceans of water are purely speculative. On the other hand, the Earth has lots of water, more than enough to cover the entire planet over a mile deep, and the Bible confirms that all the land was underwater at the time of Noah's Flood. There is plenty of physical evidence for a global flood on our planet, but evolutionists dismiss the biblical account as foolishness. For them, believing in this particular catastrophe is unscientific, despite their frequent appeals to "large impact" catastrophes in the face of conflicting evidence as we saw with the planets Mercury, Venus, and Earth.

Figure 6.27 A NASA artist's conception of Mars as it is today versus a Mars with liquid water.

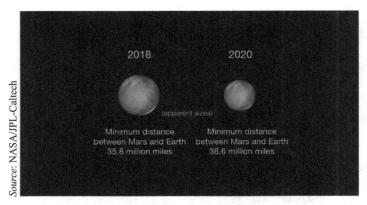

Source: NASA/JPL-Caltech

Figure 6.28 Opposition in 2018 and 2020.

Evolutionists are assuming there were oceans of water on Mars—a planet where there can be no liquid water today, while mocking creationists for believing in a watery catastrophe on Earth—a planet which has plenty of liquid water today.

Seeing Mars at its Best

The best time to view any of the planets beyond the Earth is at opposition, when they are lined up closest to Earth. Mars looks brightest (and largest in a telescope) for about one month around opposition. Unfortunately, because its orbital period is nearly twice as long as Earth's, Mars' opposition only happens an average of once every 2.1 years. Every fifteen to seventeen years, opposition occurs within a few weeks of Mars' perihelion—its closest distance from the Sun. During opposition, Mars comes relatively close to Earth, appearing seven times larger and fifty times brighter in a telescope than it does when furthest away. Opposition is also the best time to try to see the moons of Mars in a telescope.

Not all of Mars' oppositions are equal. Since its orbit is quite elliptical, some oppositions bring the planet much closer to Earth than others. Mars can appear nearly twice as large during favorable oppositions as in unfavorable ones. On August 27, 2003, Mars and Earth came as close together as they ever have—34.6 million miles—with Mars near its perihelion and Earth near its aphelion. The most unfavorable combination for opposition would be to have Mars near its aphelion and Earth near its perihelion. Our 2003 record will stand until August 28, 2287, so it really was a once-in-a-lifetime event.

6.3 Jupiter—King of the Planets

At roughly 5 AU from the Sun, the next planet we come to as we continue our outward journey across the solar system is the "King of the Planets," named for the Roman god who was supposedly in charge, Jupiter. Another name in mythology for Jupiter was Jove, so you may hear the gas giants (Jupiter, Saturn, Uranus, Neptune) being referred to as the Jovian planets. We have left the terrestrial planets behind, and now we come to planets with drastically different characteristics. In the case of Jupiter, one of the key features that stands out is quite simply its enormous size. At 1,000 times Earth's volume, you could squeeze all of the other planets neatly inside Jupiter, with its diameter about 11 times that of the Earth. At 318 times Earth's mass, Jupiter has more than double the mass of all the other planets combined. Jupiter is so massive that its gravity slightly affects the motions of the other planets. Astronomers must factor this into their computations when predicting the precise positions of planets. With its size, Jupiter is easy to see in the night sky without a telescope despite being 500 million miles away from Earth.

You may recall that Jupiter played an important role in the development of the heliocentric theory. In the winter of 1610, Galileo trained his telescope on the planet, and discovered the four largest moons of Jupiter—not orbiting the Earth as required by geocentrism, but orbiting another celestial object. Today we call these the Galilean Moons, but that is not what Galileo called them. Galileo was a shameless self-promoter, and he knew how to schmooze with royalty. The Medici family had a powerful presence in the Tuscan court, and Prince Cosimo de Medici was an important patron for Galileo. Accordingly, Galileo immediately named the objects the Medicean "stars." History suggests that this pleased the Medici family, and perhaps contributed to Galileo's overestimating how well connected he was prior to his trial.

Figure 6.29 Jupiter, the King of the Planets, with its Great Red Spot (GRS).

Like the Sun, Jupiter is composed primarily of hydrogen and helium. However, the much cooler temperature of Jupiter allows the formation of molecules such as ammonia, water, and methane from various trace elements. These molecular compounds create Jupiter's colorful cloud formations. That is what we see when we view Jupiter—clouds. Powerful winds stretch Jupiter's clouds into colorful

"belts" and "zones" that encircle the planet parallel to its equator. Belts are the dark brownish-orange features caused by gases sinking as they cool, and zones—the result of hotter gases rising from below the cloud surface—are lighter in color. Typically, the Northern and Southern Equatorial Belts that sandwich Jupiter's equator are the most visible. However, with a good telescope and a dark, clear sky, it is possible to see several thinner belts and even small disturbances taking place within them. These cloud formations are dynamic, changing from year to year, growing or shrinking and experiencing minor changes in color. A belt may even disappear completely for some time. This happened to the Southern Equatorial Belt in 2010, radically changing the appearance of Jupiter until the belt returned the following year. For any of the

Figure 6.30 A comparison of not only Jupiter and Earth, but also the Great Red Spot and Earth.

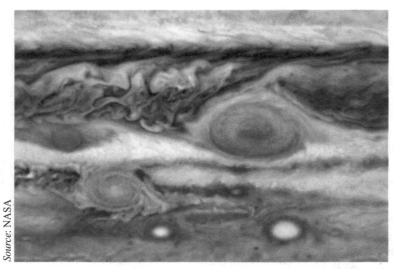

Source: NASA

Figure 6.31 The GRS, along with Red Spot Jr. and Baby Red Spot.

gas giants, belt-zone circulation transports energy by the convection of heat from the planet's interior. Relatively speaking, Jupiter is the hottest of the gas giants, so it has the most obvious belts and zones.

*In 1664, the Italian astronomer **Giovanni Cassini** saw Jupiter's banded atmosphere and a great red spot that has since been named, "the Great Red Spot." By observing how long it took the GRS to travel all the way around the planet, he was able to determine Jupiter's rotation rate—about 10½ hours for its solar day, and just under 10 hours for its sidereal day, faster than any other planet. The rotational speed is so fast, fat Jupiter bulges in the middle—an **oblate spheroid**—first seen by the seventeenth-century Dutch astronomer Christian Huygens.* While Jupiter rotates very quickly, it takes twelve Earth years to complete one revolution.

*The Great Red Spot** on Jupiter is a vast hurricane of sorts, larger than Earth, and with an estimated wind speed of 400 miles an hour. And unlike earthly hurricanes, the GRS doesn't die down; it's still going strong hundreds of years since its discovery. The GRS has grown and shrunk over time, and even changed colors from blue to red to orange, and occasionally other red spots have developed and sometimes even been absorbed by GRS!* Since Jupiter rotates, the Great Red Spot is out of sight on the other side of the planet half of the time. But even when on the Earth-facing side of Jupiter, the Great Red Spot is hard to see near the limb (the edge of the disk). The best view is when the spot is said to "transit" or pass across the face of the disk, an occurrence predictable up to a month or so in advance. Backyard astronomers can locate future transit times posted on various astronomy websites.

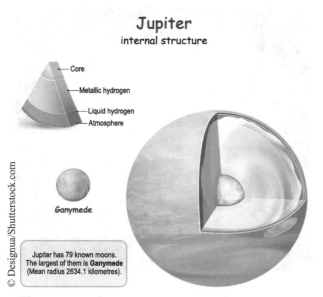

Jupiter
internal structure

- Core
- Metallic hydrogen
- Liquid hydrogen
- Atmosphere

Ganymede

Jupiter has 79 known moons. The largest of them is **Ganymede** (Mean radius 2634.1 kilometres).

© Designua/Shutterstock.com

Figure 6.32 One proposed model for Jupiter's interior.

Jupiter is a great big ball of hydrogen and helium gas. But one cannot simply plunge into it and expect to come out the other side. The gas pressure increases with depth, eventually becoming liquid. *The deeper we go, the thicker and hotter the air becomes, until at the core of the planet, where the pressures are so great—50 million Earth atmospheres—that hydrogen is compressed into a liquid metal, and the temperature is four times hotter than the surface of the Sun!* There's so much of this exotic liquid metallic hydrogen inside Jupiter that it transforms the planet into an enormous generator. The deep layer of liquid metallic hydrogen combined with Jupiter's rapid rotation is thought to create a magnetic field 450 million miles long—the biggest entity in the solar system. Jupiter's magnetic field is thousands of times greater than that of Earth. Jupiter's magnetosphere can produce up to 10 million amps of electric current.

The magnetosphere deflects the solar wind and captures all sorts of high-energy particles. Since the magnetosphere is so much stronger than Earth's, it traps and holds more particles and higher-energy particles. The result is levels of radiation around Jupiter that would be lethal to humans.

*Like Earth, Jupiter boasts **auroras** at its poles. However, these shimmering halos are a thousand times stronger than any Northern Lights display found on Earth.* On Earth, an aurora is produced by the interaction of the solar wind with our magnetosphere, but on Jupiter the planet's own magnetosphere is powerful enough to produce an aurora without the solar wind. *Tremendous lightning storms flash and rumble through Jupiter's sky, dwarfing any comparable thunderstorm activity ever experienced on our planet.*

Observations of the magnetosphere are consistent with an estimated biblical age of about 6,000 years but are difficult to explain with the solar nebular model and the assumed age of 4.6 billion years. For example, since magnetic fields naturally decay with time, it is hard to understand how Jupiter could maintain such an incredibly powerful field for over 4 billion years. Another indication of youth is Jupiter's internal heat. Jupiter emits nearly twice the amount of energy that it

Figure 6.33 Aurora activity near Jupiter's poles.

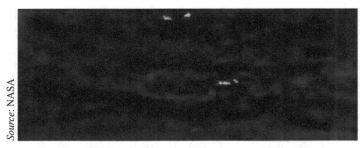

Source: NASA

Figure 6.34 Lightning discharges in Jupiter cloud tops.

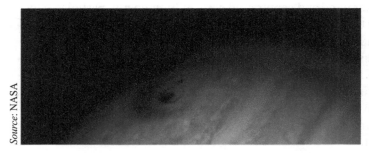

Source: NASA

Figure 6.35 Comet SL9 impacts on Jupiter.

receives from the Sun. Planets gradually cool as they radiate heat into space, and Jupiter is large enough to do this for thousands of years. But if it were really billions of years old, why hasn't it cooled off by now?

*Back in the summer of 1994, something amazing happened—a comet named **Shoemaker-Levy 9**, broken into over twenty pieces by Jupiter's immense gravity, struck the giant planet! Comet debris was scattered throughout the Jovian atmosphere.* In 1995, the spacecraft *Galileo* arrived in time to survey the damage and study the atmosphere. *The largest impact gave Jupiter a black eye twice the size of Earth; the explosion was equal to that of 6 trillion tons of TNT, the same amount of energy given off if you could explode an atom bomb once a second for ten years. Had this comet hit our planet, it would have been a bad day for Earth. But Jupiter just kept right on rolling along, incorporating the comet material into itself. And we have seen other, smaller impacts on Jupiter since then!*

Comparing Predictions with Observations

The *Galileo* spacecraft made observations that conflicted with predictions of the solar nebular model. For example, it turned out that Jupiter had relatively large quantities of the noble gases argon, krypton, and xenon. This was not possible given Jupiter's location and the tenets of the nebular theory. In an earlier chapter, we already discussed problems with accretion and planetesimals and the notion that "dust" grains can produce asteroid-sized bodies. As a paper in Nature pointed out, "how this process continues from meter-sized boulders to kilometer-scale planetesimals is a major unsolved problem." The formation of planetesimals is a problem for any of the planets, but even if you could make planetesimals from gas and dust, secular scientists couldn't ignore or explain why Jupiter's chemical composition didn't match their models. According to secular models, Jupiter can't have these elements in the concentrations that we observe because those elements can't have been in Jupiter's region of the solar system back when Jupiter was forming. According to the models, the only place that these elements are allowed to have formed is way out beyond the orbit of Neptune, more than 3 billion miles from where Jupiter is today.

To try and solve this problem some evolutionists have suggested that Jupiter must have formed way out there and then somehow moved inward. However, the secular model says there wouldn't have been enough material out there for Jupiter to form, and this has been supported by current observations of exoplanets and disks we see around other stars. Others have suggested that the unexpected elements came from planetesimals or asteroids that formed way out there and then moved billions of miles inward, delivering these elements to Jupiter but that doesn't work either. If they had formed out there and then moved in they would have warmed up as they moved toward the Sun and as they got warmer they would have lost whatever argon, krypton, and xenon they contained. Recently, some theorists have been going the other direction altogether, saying Jupiter formed *closer* to the Sun than its position today, and eventually (somehow) migrated outward rather than inward (the "Grand Tack" hypothesis), creating a new set of problems. Based on the nebular model, Jupiter shouldn't be where it is, with the composition it has. *Galileo's* observations created big problems that have not been solved in the secular framework.

Moons (and Rings?)

Until 2003, when it was purposely "retired" (a forced crash) the *Galileo* spacecraft also investigated the larger moons of Jupiter, giving us far more observational evidence to work with. Jupiter has at least sixty-seven moons, but based on confirmed satellites and proposed recent finds, it is safe to say that number is higher (over seventy). Most of these moons are unimpressive, often with diameters of less than 10 miles. Some orbit Jupiter in the wrong (retrograde) direction and have tilted, elliptical orbits. However, it is the Galilean satellites that steal the show and capture our attention.

In order of increasing orbital distance from Jupiter, these satellites are Io, Europa, Ganymede, and Callisto, named after various mythological love interests of Jupiter. These moons are solid bodies composed of rock and ice. Io takes only 1.77 days to orbit Jupiter, and the orbital periods of Europa, Ganymede, and Callisto are 3.55 days, 7.15 days, and 16.7 days, respectively, just as Kepler and Newton would have predicted. There is a pattern here. Europa's period is twice that of Io, and Ganymede's period is twice that of Europa. This is an example of *orbital resonance*. Callisto breaks from the pattern,

Figure 6.36 A combination of images from the *Galileo* spacecraft giving size comparisons with Jupiter's Great Red Spot.

Image courtesy of NASA

and is not in resonance. In terms of size, Europa is a little smaller than Earth's Moon, and Io is just a bit bigger. Callisto and Ganymede are about 40 percent and 50 percent larger than our Moon, respectively. In fact, Ganymede is the largest satellite in the solar system, dwarfing not only our Moon but also the planet Mercury.

Compared to Jupiter, Io is tiny, but it's also one of the most spectacular places in the solar system. High-resolution images of Io taken by the *Voyager* spacecraft in 1979 revealed that, unlike all other known moons, Io has no impact craters. It is covered with sulfur compounds that are responsible for its colorful surface which is covered with volcanoes. Io has at least 400 volcanoes, and at least 150 of them are active today. "Loki," for example, is more powerful than all the Earth's volcanoes combined. These volcanoes can become taller than Mount Everest. Eruptions are nearly constant, covering the surface with volcanic material and erasing any previous record of impacts. Io is the most volcanically active world in the solar system. When *Voyager 1* flew past Io, it detected nine volcanoes erupting simultaneously! The "gravitational stretching" occurring every 3.55 days, when Io passes between Jupiter and Europa, seems to provide some of the internal energy required for Io's volcanism. The question becomes whether or not this activity could continue over billions of years.

Little Io puts out twice as much heat as the Earth does. Some of it comes from tidal flexing in a gravitational tug of war between Jupiter on the one side and Jupiter's other large moons on the other. They're both pulling on Io and this squeezing and flexing generates heat in Io's interior. The problem is that this squeezing can only account for some of Io's heat. Remember, "old" objects have lost their own heat, while "young" objects still have their own heat. If Io is really billions of years old, that additional energy would have dissipated long ago. In addition, we have measured the lavas coming from the volcanoes and (1) there's an amazing amount of it, (2) it has extremely high temperatures, and (3) it has very high density. This does not seem consistent with an object that is billions of years old. If Io really was billions of years old, these temperatures and densities shouldn't be possible. After the first few million years Io should have formed a low-density crust

Figure 6.37 Io.

Figure 6.38 Volcanic eruption on Io.

after all the higher-density material sank down to the inside but there are still high-density materials on the surface—Io looks quite young.

Europa, perhaps the smoothest of all moons, is covered with a shell of shiny, clean ice, which even secular astronomers refer to as "young." The ice is several miles thick but some scientists believe there might be an ocean of liquid water beneath it. For this reason, Europa is high on the list of places NASA would like to search for life, promoting the notion that where there is water, there is life. Of course, that is not necessarily true, but it helps with funding. Whether it is Mars, Europa, or any of the other hopeful targets in the search for life, the search is driven by a simple set of assumptions. (1) All there has ever been at work in the universe is natural law, (2) somehow natural law gave rise to life on Earth from non-living chemicals over billions of years, and (3) since nature did it here, it should do it elsewhere. Note that none of these are scientific statements. Rather, these are the philosophical assumptions that arise from naturalism.

Galileo's trips past Europa also raised important questions about the evolutionary model for cratering. We've seen that evolutionists like to use craters to date things in the solar system, arguing that the more craters something has, the longer it's been sitting there getting struck by other objects. It previously was thought that most of the craters seen on moons and planets were the work of direct impacts from asteroids and comets. *Galileo's* observations suggest that most of those craters might instead be "secondaries," impacts that formed by the material ejected from an initial impact. As reported in *Nature*, Europa's secondaries account for as much as 95 percent of all the small craters observed on the moon. This finding has clear implications for how astronomers date the ages of planetary and satellite surfaces. Planets and moons have been struck with a lot fewer impacts than evolutionists thought, calling into question the billions of years that these objects have supposedly been around.

Ganymede has one of the most bizarre surfaces in our entire solar system. It's crisscrossed with ridges and "grooved terrain" (the actual term) but with weird patterns streaking across it all. Some

Figure 6.39 Europa.

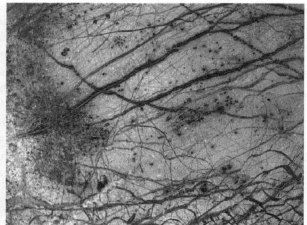

Figure 6.40 Europa's icy surface.

© mr. Timmi/Shutterstock.com

Figure 6.41 Ganymede.

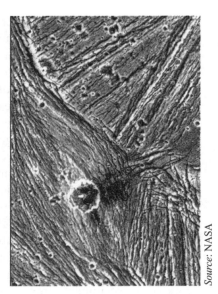

Source: NASA

Figure 6.42 Ganymede close-up.

places are rough and rocky, other places are flat and smooth. In 1996, *Galileo* made the surprising discovery that Ganymede had its own magnetic field. This is a challenge for people who try to explain how it formed without a Creator. Evolutionary models predicted that Ganymede couldn't have a magnetic field but the observations did not match the predictions of the model. However, since Ganymede does have a magnetic field it is a moon that has its own aurorae. NASA's Hubble Space Telescope was able to provide images of one such event. As one secular college textbook noted, "the cause of Ganymede's unique magnetic field remains a puzzle."

Recall that Callisto was not in orbital resonance as were the other Galilean moons. Callisto is far enough from Jupiter that it escapes some of the more severe gravitational effects, avoiding the "squeeze" that the other three moons experience to varying degrees. Callisto is the most heavily cratered object in the solar system. Evolutionists believe that this moon has one of the oldest surfaces of any object, at about 4 billion years. It was a real surprise then when our space probes took some close-up pictures. Evolutionists expected a large number of smaller craters to go with the big ones but they aren't there. They should be there if the surface is really billions of years old. Callisto doesn't match the predictions of the evolutionary cratering model. Not only that, some of the pictures show what appears to be fresh ice on Callisto's surface. This suggests that there is still erosion and geological activity going on, contrary to evolutionary models and predictions. Like Europa, Callisto draws special attention from NASA, as it too may have water beneath its crust. Based on how Callisto interacts with Jupiter's magnetic field, some scientists speculate that there must be a significant amount of salt water somewhere under the satellite's surface.

Jupiter also has a system of tenuous rings that were first detected in 1979 by the *Voyager 1* spacecraft. These thin rings are much less substantial than Saturn's, and too faint to be seen with most

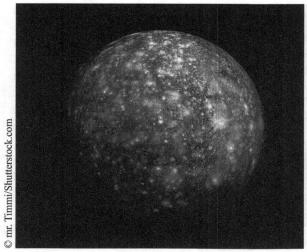

Figure 6.43 Callisto.

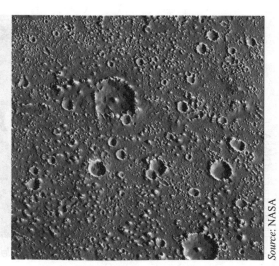

Figure 6.44 Lots of craters!

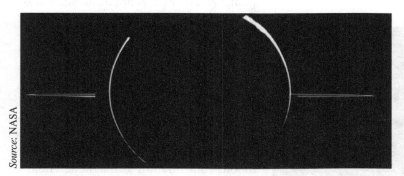

Figure 6.45 Night side of Jupiter, showing its thin ring.

Earthbound telescopes. The main rings consist of dust particles orbiting mainly around and inside the orbit of Jupiter's innermost moons, Metis and Adrastea, with an even thinner set of rings called the "gossamer rings" extending further out. All of the gas giants have some sort of faint rings, but there is nothing faint about Saturn's rings.

6.4 Saturn: Lord of the Rings

Travel nearly 10 AU from the Sun—roughly double the distance from the Sun to Jupiter, and we find what may be the most remarkable planet in our solar system when it comes to looking gorgeous from a distance. Saturn brings the "wow" factor, and anyone who has ever seen it for themselves in a telescope knows exactly what that means (not that there aren't plenty of stunning photographs). In Roman mythology, Saturn was associated with agricultural bounty and in some myths served as the god of time and the father of Jupiter. The ancients would have been stunned to know what the planet really looks like. Naked eye, it is simply a disk of reflected light just like every other planet. *It's almost a billion miles from Earth, depending on its orbital position—a distance so great that only a handful of spacecraft have made the journey: *Pioneers 10* and *11, Voyagers 1* and *2, and the New Horizons* probe to Pluto—all of which had brief encounters with Saturn as they flew past it; and another spacecraft named

Figure 6.46 One of many beautiful *Hubble* images of Saturn.

Image courtesy of NASA

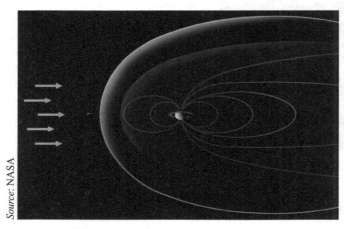

Figure 6.47 Saturn's magnetosphere.

Source: NASA

Cassini, a workhorse that arrived in 2004 and retired (crashed) in 2017. Saturn is nine times wider than Earth. But its average density is less than water. Given an ocean large enough, Saturn could float!* Saturn rotates about once every 10.5 hours—its sidereal day. This high-speed spin causes Saturn to bulge at its equator and flatten at its poles, resulting in a shape we call an oblate spheroid. The planet is around 75,000 miles across at its equator, and 68,000 miles from pole to pole.

Saturn's magnetic field is hundreds of times stronger than Earth's. Within Saturn's enormous magnetosphere, the planet and its moons are influenced more by the magnetic field than by the solar wind. Aurorae occur when charged particles spiral into a planet's atmosphere along magnetic field lines. For Earth's aurorae, the charged particles come from the solar wind. *Cassini* showed that at least some of Saturn's aurorae are like Jupiter's and are caused by a combination of particles ejected from Saturn's moons and Saturn's magnetic field's rapid rotation rate rather than the influence of the solar wind. However, these "non-solar-originating" aurorae are not completely understood yet.

Saturn's magnetic field doesn't match evolutionary predictions. Saturn dumbfounded planetary theorists who study dynamo models by having a highly symmetric internal magnetic field. A field that is symmetric about the rotation axis violates a basic theorem of magnetic dynamos. Saturn's magnetic field doesn't seem to be coming from a dynamo. This would be impossible if Saturn were billions of years old, but there's no problem at all if Saturn is only thousands of years old. As we would read in a current, secular textbook, the alignment of Saturn's magnetic field "is peculiar, is not observed for any other planet, and is not understood."

*Space is a vacuum, and sounds cannot be heard in a vacuum. But scattered throughout space are tiny grains of dust, ice crystals, magnetic fields of charged particles . . . and when the *Voyager 1* spacecraft crossed Saturn's bow shock, that region where the magnetic field of a planet (the **magnetosphere**) encounters the solar wind, its microphone picked up the interplay of subatomic particles as *Voyager* plunged through their midst. Besides measuring the powerful magnetospheres surrounding the giant planets both *Voyager* spacecraft recorded the sounds of ice crystals striking their hulls as they crossed Saturn's rings.*

Figure 6.48 Ammonia ice storm on Saturn.

Saturn is a slightly smaller version of Jupiter in some respects. As a fellow gas giant, it is made of hydrogen and helium and trace amounts of molecules such as the methane and ammonia that give its yellow color. As with Jupiter, colorful clouds are stretched into belts (dark-colored) and zones (light-colored). However, Saturn's belts and zones are more subtle than Jupiter's. *Saturn is colder than Jupiter, so its clouds form lower in the atmosphere and are hidden by a high altitude haze, giving the ringed planet a dull, bland appearance compared to the candy-ribboned clouds of Jupiter.* The yellow and gold bands seen in Saturn's atmosphere are the result of superfast winds in the upper atmosphere, which can reach up to 1,100 mph around its equator, combined with heat rising from the planet's interior, as Saturn radiates away more energy than it receives from the Sun. *Under false color imaging spectacular atmospheric details can be seen, such as immense, planet-wide ammonia ice storms, many times larger than the Earth.*

At Saturn's north pole, a strange, hexagonal storm, over 15,000 miles across, slowly turns with the planet's rotation. We can think of this as a six-sided jet stream. This pattern was first noticed in images from the Voyager I spacecraft and was then more closely observed by the Cassini spacecraft. The hexagon is a wavy jet stream of 200-mph winds with a massive, rotating storm at the center. There is no weather feature like it anywhere else in the solar system.

How did Saturn form? Recall from earlier in this chapter that based on the solar nebula model, Jupiter wouldn't be here today if the model was true, and the same can be said for Saturn. Gravitational interaction between the gaseous protoplanetary disk and the mass of planetary cores causes them to move rapidly inward over about a hundred thousand years in what we call the migration

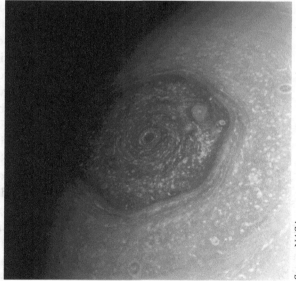

Figure 6.49 Weird, hexagonal storm.

of the planet. In the disk, theories predict that the giant protoplanets will merge into the central star before planets have time to form. This makes it very difficult to understand how they can form at all. Astronomers call this the "migration problem." In simple terms, Jupiter and Saturn would have both migrated inward and slammed into the Sun billions of years ago if the nebular hypothesis was true. One report concluded this way: "understanding the formation of giant planets is currently one of the major challenges for astronomers." If the evolutionary model says that Jupiter and Saturn shouldn't be there but they are there, that is a major challenge indeed. That is not to say that astronomers do not speculate on new ideas to try to "save the appearances," but like Ptolemy's epicycles the need for so many new ideas may speak to an issue with the overarching model.

Seeing Rings

Galileo could not clearly discern the rings in his early telescope. In 1610, he noted bulging "companions" on either side of the planet, and then was astounded in 1612 when those structures seemed to vanish. It would not be until 1656 that Dutch astronomer and lens maker Christiaan Huygens correctly identified the rings, publishing a full accounting of his study of Saturn in *Systema Saturnium* in 1659. In 1675, Jean-Dominique Cassini, the director of the Paris Observatory, discovered that the rings of Saturn were not a single entity as proposed by Huygens, but had a gap of some sort separating the rings into two concentric bodies, the outer designated as "A" and the inner as "B." This 3,000-mile wide gap became known as Cassini's Division. As the centuries passed, more "gaps" were discovered, adding C, D, E, F, and G. Today, astronomers believe that what causes and maintains these gaps are the interactions and resonances with the moons.

© Morphart Creation/Shutterstock.com

Figure 6.50 Christiaan Huygens.

Saturn has a tilted axis, much like Earth at about 27 degrees of inclination. Saturn's rings orbit its equator. The effect then is that sometimes we see the rings "facing" us while at other times the rings are visibly on edge and practically disappear for a time. Approximately every fifteen years— twice in each 29.5 Earth year period of revolution, the rings seem to turn edgewise to observers on Earth. This coincides with Saturn's equinoxes, the times when Saturn's equatorial plane intersects the Sun. *When this happens, the thin edge is simply too dim to be visible when seen against the bright disk of Saturn. At these times Saturn appears as a small, featureless, yellow ball, and not very impressive. But with the loss of the interfering bright light of the rings it's possible to see other, fainter objects near the planet, such as small moons, and many of Saturn's satellites have been discovered when the rings go edge-on.*

While astronomers were discovering the gaps of the ring system, the question remained about the make-up of the rings. Huygens thought they were solid, Cassini showed there was a gap and subsequent gaps were added, but what were the rings made of? There was plenty of speculation, but no real proof. In 1856, a young James Clerk Maxwell tackled the question using differential calculus to model Saturn and its rings. He showed that a solid ring or rings could not be stable. He also showed that a fluid ring could not be stable and would break up into small portions. Maxwell then examined a single ring composed of unconnected satellites—"moon-

Saturn Ring-Plane Crossing
Hubble Space Telescope · Wide Field Planetary Camera 2
PRC96-16 · ST ScI OPO · April 24, 1996 · E. Karkoschka (LPL) and NASA

Source: NASA/JPL/STScI

Figure 6.51 Saturn's rings nearly disappear in the upper image when Earth is in the plane of the rings.

lets" if you will. Maxwell concluded that the rings were made of "unconnected particles." To give us the appearance of a solid ring, each ring would have to be made up of a very large number of independent particles.

Today, Spectroscopes and radio signals passing through the rings tell us that there are billions of ice particles, ranging in size from tiny grains and pellets to icebergs as big as a house, all traveling in their own separate orbits of Saturn. The largest chunks are found among the inner rings, and the smallest ice bits are farthest out. This "carpet" of ice moonlets and particles stretches about 175,000 miles from edge to edge (diameter), and about 40,000 miles out from the planet to the outer edge of the major rings. Yet this carpet is remarkably thin—just a few hundred feet thick in the thicker areas—less than the length of a football field! If you tried to build a model of Saturn using a basketball, the rings would be thinner than an edge of a sheet of paper.

So how did the rings get there? Creationists believe they were cre-

Cassini's Gap

A ring
B ring
C ring
Enke's Division

Source: NASA—labels by Jon Bell

Figure 6.52 The Rings of Saturn, labeled.

ated at the same time Saturn was, thousands of years ago. Secular astronomers could never accept this. While most astronomers acknowledge that they aren't sure about how the rings formed, there is no shortage of speculation. You might anticipate one proposal—an impact hypothesis. This model suggests that an asteroid hit one of Saturn's satellites and smashed it into pieces, creating the material for the rings. Another hypothesis suggests that a moon-sized object from the outer solar system flew near Saturn, where tidal (gravitational) forces ripped it apart. There is a report from NASA on Saturn's rings that simply says "Saturn's rings: age and origin unknown."

Marvelous Moons

Like Jupiter, Saturn has many satellites (perhaps more than Jupiter?) in a wide range of sizes. The first moon of Saturn was discovered by Huygens in 1655 when he located the largest of the satellites, Titan. By 1847 there were seven known moons, and astronomer John Herschel started naming them after characters from mythology. The number of moons orbiting Saturn (not counting the moonlets of the rings) is a moving target. There are named, confirmed discoveries but also dozens more to be confirmed. It is safe to say there are at least sixty (fifty-three are named), but with the additional moons to be confirmed, Saturn may exceed Jupiter's count. *Most are quite small—ranging in size from just a few miles across to several hundred miles wide. Only 5 are larger than 500 miles across. They're mainly made of water ice, covered by a thin layer of dust. *Cassini* made startling discoveries about many of them:

Iapetus sports a 15 mile-high ridge (dubbed, "the belly band") that partly encircles the Moon at the equator. A dark, black smudge coats the leading edge of Iapetus, possibly dark carbon dust dumped onto it by Saturn's outermost moon, Phoebe.*

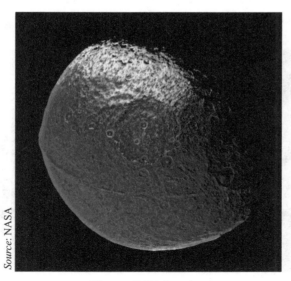

Source: NASA

Figure 6.53 Iapetus.

Enceladus is bright white and has only small craters. *Cassini* discovered plumes of icy material being ejected like geysers near this little moon's south pole, indicating that Enceladus has significant internal heat. This is problematic for the view that the solar system is billions of years old because if the Moon were really that old, that internal energy should have escaped long ago, and Enceladus does not experience enough gravitational (tidal) tugging to regenerate that heat. If we took a close look at some of the other moons of Saturn we would find that Enceladus' neighbors are much brighter than they're supposed to be. Apparently, Enceladus is spray painting them with ice and snow. Evolutionary models say that Enceladus should be old, cold, and dead but it looks like it's none of those things. Enceladus is evidence for a young solar system.

Figure 6.54 Enceladus.

Figure 6.55 Ice jets near south pole.

Mimas orbits close to the planet and is easy to recognize because it has an enormous crater on one side, with a central peak about the size of Mount Everest. The crater (named Herschel) is about one-third the diameter of Mimas and makes the moon look a bit like the Death Star from *Star Wars*. Mimas helps to maintain the Cassini Division, orbiting outside the main rings in a 1:2 resonance with any moonlet that might wander into the Cassini Division. Mimas orbits once every time a Cassini moonlet orbits twice. Moonlets would repeatedly get a gravitational tug at the same point in their orbit, leaving the Cassini Division clear of wayward moonlets.

Figure 6.56 Mimas.

Figure 6.57 Herschel crater on Mimas.

Saturn is the only planet known to have moons that share a common orbit at precisely the same speed so they never collide. We call these objects Trojan moons—moons that share an orbit with another, larger moon. Trojan moons are found near gravitationally stable points ahead or behind a larger moon. These locations, called *Lagrangian points*, are 60 degrees ahead of the large moon in its orbit and 60 degrees behind, respectively. The larger moon, smaller moon, and planet form an equilateral triangle—a very stable configuration. Saturn's moon Tethys shares its orbit with Telesto (leading Tethys by 60 degrees) and Calypso (which trails Tethys by 60 degrees), so Telesto and Calypso are "the Tethys Trojans." Likewise, the larger moon Dione shares its orbit with Helene (which leads) and Polydeuces (which follows).

Titan is the second-largest moon in the solar system (behind Jupiter's Ganymede), with a diameter of about 3,200 miles. *It is larger than the planet Mercury, and it may have a rocky core.* Titan is the only moon we have found with a thick atmosphere. This atmosphere is composed primarily of nitrogen with traces of methane and other hydrocarbons. These trace molecules give rise to Titan's orange color and make it nearly impossible to see any surface features. The existence of methane in Titan's atmosphere challenges the notion that this moon is billions of years old because ultraviolet radiation from the Sun would break down the methane in a timescale of only 10 or 20 million years. However, there is still plenty of methane there today. If Titan was really billions of years old it would require two things, (1) a source of methane to keep replenishing the atmosphere, and (2) a lot of ethane built up on its surface from all the methane that was broken down. The *Cassini* spacecraft carried a special probe (aptly named "*Huygens*") to drop onto Titan and investigate this mysterious moon.

Source: NASA

Figure 6.58 Titan, seen in infrared.

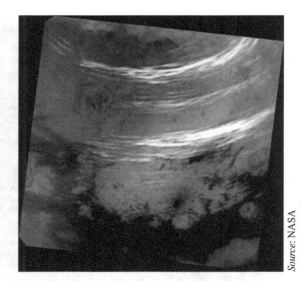

Source: NASA

Figure 6.59 Methane clouds on Titan.

Source: NASA

Figure 6.60 Titan's surface.

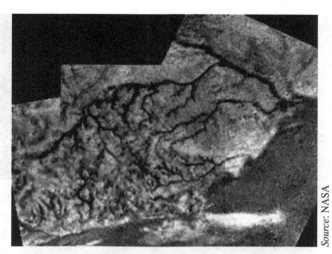

Source: NASA

Figure 6.61 Nitrogen river valley.

Based on a presupposition of billions of years, secular scientists confidently predicted we would find a deep global ocean of methane and ethane on the cold surface of Titan. One scientific paper predicted a layer of hydrocarbons about a half mile deep. One of the leaders of the Titan project went even further, predicting that this remarkable moon could be covered by a global ocean of ethane with an average depth of well over a mile. When *Huygens* finally penetrated the haze, the truth was confirmed. Despite evolutionary expectations, there is no global ocean of methane and ethane on Titan. Titan's surface is relatively dry. It appears that there are a few scattered lakes of methane and ethane (with proportionally less ethane than expected) but nothing close to the predictions. When you combine this with the general rarity of craters on Titan, the Moon does not fit the predictions for an object billions of years old. Titan looks young, and what we observe fits with an ages of just thousands of years.

*Among the rings of Saturn there are also some very small moons, called **shepherding satellites.** Their gravitational influences can keep some of the rings together, preventing them from falling in toward Saturn or escaping outward. They can also put "kinks" or "braids" in the rings.* For example, Prometheus and Pandora are two small moons responsible for the thin F-ring. Prometheus orbits just inside the F-ring, and Pandora orbits just outside. These are called *shepherd moons* because they gravitationally deflect any wayward moonlets back into the F-ring.

Janus and Epimetheus are known as "the dancing moons of Saturn." In their orbital ballet, Janus and Epimetheus swap positions every four years—one moon moving closer to Saturn, the other moving farther away. During the switch, the difference between the two orbits of these moons is *less* than the diameter of either moon. How can Epimetheus pass Janus (or vice versa) without collision? These moons have been observed for many decades, and a crash has yet to happen.

Source: NASA

Figure 6.62 Rhea and other moons appear near Saturn's rings.

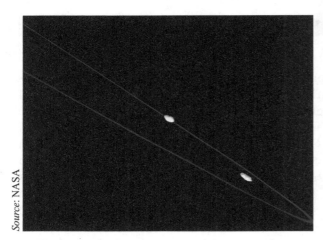

Source: NASA

Figure 6.63 Pandora and Prometheus shape Saturn's F-ring.

It turns out that the mutual gravity of the two moons causes them to switch their orbits rather than colliding. Right before a potential collision, their gravitational interactions will cause one moon to get a small "kick" of energy to switch to a higher orbit while the other experiences a small loss of energy and falls to a lower orbit. Four years later, it all happens again and they trade places.

As we complete this section on Saturn, and this chapter on four planets with such diverse features, consider the artistry, creativity, and design of our solar system. The Heavens really do declare the glory of God. What an amazing solar system the Lord has created!

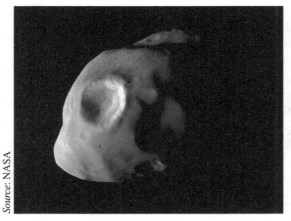

Source: NASA

Figure 6.64 Pandora.

Source: NASA

Figure 6.65 Prometheus.

That is obvious to anyone whose worldview is based on biblical teaching. However, naturalism is just one example of how mankind attempts to suppress the truth, as we read in Romans Chapter 1. Even the most radical atheist will look at a clear night sky and experience awe and wonder, but rather than hearing the declaration of the

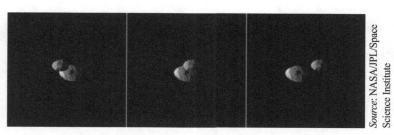

Source: NASA/JPL/Space Science Institute

Figure 6.66 *Cassini* documented Epimetheus passing Janus, with each photo about one minute apart.

glory of God, they will find glory in chance, time, the Big Bang . . . anything except a Creator. Once truth is suppressed—the Truth about who God is and what He has done—then we are left with cheap substitutes.

The Solar System—Part 3

Chapter 7

> *"A true scientist cannot be an atheist. When you peer so deeply into God's workshop and have so many opportunities to marvel at His omniscience and eternal order, as we have, then you should humbly bend your knees before the throne of the most holy God."*

—**19th century astronomer Johann Madler**

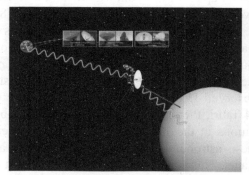

Figure 7.1 Voyager 2 spacecraft.

Figure 7.2 New Horizons spacecraft.

Figure 7.3 Dawn spacecraft.

Images courtesy of NASA

Figure 7.4 William Herschel.

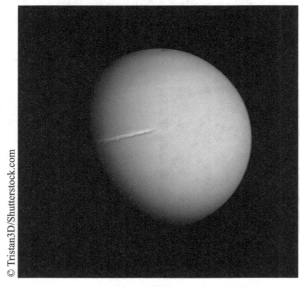

Figure 7.5 Uranus.

7.1 Uranus
The Musician Who Found a Planet

From the beginning, the objects we now know as Mercury, Venus, Mars, Jupiter, and Saturn were visible to everyone as the peculiar wanderers of the night sky. We do not discuss them in terms of being "discovered." With the advent of the telescope in the early 1600s, we quickly learned more about these five. However, we don't find a lot of information to suggest that early astronomers were actively searching for additional planets in our solar system. Perhaps some astronomers had no reason to entertain the possibility of more planets, while others had telescopes that simply weren't up to the task, but regardless, the first discovery of a new planet did not occur until 1781.

With the benefit of hindsight, it turns out that the new planet had been seen and mapped multiple times prior to 1781, but had been misidentified as a star. In 1690, John Flamsteed observed Uranus and cataloged it as the star 34 Tauri. Pierre Lemonnier made similar observations in the 1750s and 1760s, but neither of these astronomers recognized the object as a planet. This is understandable as Uranus is a tiny disk and moves very slowly compared to the other planets. The planet is visible with binoculars but remains indistinguishable from the stars at such low magnification. Technically, Uranus is considered to be a naked eye object, but in practice it is highly unlikely for anyone to recognize it without assistance even in the best of conditions.

William Herschel was a church organist in Bath, England. He also composed music, and the first performance of Handel's oratorio *Messiah* in Bath was conducted by him. He had a great interest in the sky and astronomy, and in telescopes. (During intermissions, he would go outside and look at the stars.) But then (as now), most musicians don't make much money, and telescopes were amazingly expensive. So Herschel did the economical thing and made his own. It's painstaking work to make your own mirrors, and Herschel did it from scratch forming and shaping the glass slabs, then spending hours upon hours grinding the glass to just the right curve and smoothness. If you stop while polishing the mirror, you're apt to introduce flaws. Herschel's very dedicated sister (and ultimately an accomplished astronomer in her own right) **Caroline used to break up morsels of food and hand-feed him so he wouldn't have to stop polishing to eat.**

With Caroline's assistance, he systematically cataloged thousands of "deep sky" objects (lying beyond the solar system). His survey initiated the development of the *New General Catalogue (NGC)* of celestial deep sky objects that is still used by astronomers today. Herschel also specialized in observations of binary stars. It was during this cataloging of binary stars that Herschel observed a small, light-blue disk. It could not have been a star since stars are too distant to appear as disks. Even in a telescope they appear as points of light. However, this object appeared as a small, distinct sphere. Hershel initially presumed that he and Caroline may have discovered a comet. But as he tracked the object over several nights, he found that it did not move like a comet. Comets generally have highly elliptical orbits. They also have distinctive structures such as their iconic tails, and this object had none of those features. Could it be a planet?

When William sent a report of his discovery to the Royal Society in London, he was still the organist in Bath, and did not

Figure 7.6 Replica of the telescope that William Herschel built, and that he and Caroline then used.

Image courtesy of NASA

attend the meeting himself. Once the report was reviewed, excited astronomers left London and headed to Bath to see Herschel's basement workshop and his garden observatory. (The home is a museum today.) Shortly after this initial excitement, the planet (in Caroline's words) "took a summer holiday." It hid in the daytime sky and delayed more detailed study needed for final confirmation for several months. Once it reappeared, astronomers all across Europe carefully followed it, watching for typical comet behavior—a parabolic path, a steady change in brightness, a tail–none of which occurred. Instead, the observations showed an object double the distance to Saturn, orbiting the Sun in a nearly circular path. This was a planet. As word spread that William Herschel had single-handedly doubled the width of the solar system, this amateur astronomer/musician was recognized by the King of England and given a royal appointment and a meaningful salary increase.

Our seventh planet was nearly named "George." Keep in mind that historically, England had just lost control of the American colonies—a disastrous result for King George. This same king had shown Herschel a great kindness by making him a professional astronomer, and in return Herschel named the new planet "Georgium Sidus" (George's Star). This did not sit well with traditionalists who wanted the name of a Roman god, or with other nations who weren't fond of the king. The French promoted the idea of calling the object "Herschel" (or Herschel's Star), and did so for several decades. The Brits fell into the habit of calling it "the Georgian" (George). However, much of Europe followed the lead of German astronomer Johann Bode, the director of the Berlin Observatory, in the push to name the new planet Uranus (your-un-us), after the mythological god of the sky. During this time of uncertainty about the name, the Germans doubled down on their proposed name when another German, named Martin Klaproth, took the liberty of naming a new element "Uranium" in honor of the planet in 1789, and of course eventually Uranus became universally accepted.

Uranus, the Sideways Planet

Uranus orbits the Sun at an average distance of about 1.8 billion miles—nearly 20 times farther out than Earth. Accordingly, Uranus takes eighty-four years to orbit the Sun. The planet is four times the diameter of Earth, and by volume it could hold about sixty-four Earths. As one of the four gas giants, its outer composition is similar to that of Jupiter and Saturn—mostly hydrogen and helium gas, with a small percentage of methane, which gives the lovely blue-green color by reflecting the light from that end of the spectrum, while absorbing colors near the red end of the spectrum. Based on its density and spectral information of its atmosphere, the interior of Uranus is thought to be composed of various ices such as water, ammonia, and methane. For this reason, Uranus is sometimes referred to as an *ice giant*.

Recall that all the gas giants have some sort of ring system. The system of rings around Uranus is quite different from the rings of Saturn. Saturn's main rings are in broad sheets, while the rings of Uranus are more like a series of thin ropes. Each encircles Uranus at a discrete distance, all in the plane of its equator. These rings were discovered in 1977 when astronomers observed Uranus passing in front of a bright star. Surprisingly, the star "winked" five times immediately before Uranus

Figure 7.7 The tilt of Uranus puts it on its side, rolling like a bowling ball.

passed in front of it and again five times immediately afterward. They correctly deduced that a system of five narrow rings surrounds Uranus. Eight additional, more tenuous rings were detected at later dates, including two sets of rings first spotted by *Voyager*. Since their initial discovery, Uranus' rings have been imaged directly by the *Voyager 2* spacecraft and also by the Hubble Space Telescope. Because these rings appear to be stable, narrow, and well defined, many astronomers have concluded that there are additional small moons that act as shepherding satellites.

A riddle for astronomers is why all of the gas giants radiate more energy into space than they receive from the Sun except for Uranus. Neptune, practically a twin to Uranus in size and structure, radiates more than 2.5 times the energy it receives. Why doesn't Uranus? These planets all supposedly formed at the same time from the same materials by the same natural processes. That means they should have turned out roughly the same, but Uranus is very different than the others in this important area. Perhaps the bigger question is why the other three gas giants do radiate so much energy if they are billions of years old, but the difference here is striking.

Given the great distance from the Earth to Uranus, detailed study of the planet was limited until the *Voyager 2* spacecraft visited in 1986, making a variety of observations and providing images of a nearly featureless blue sphere without any prominent belts and zones, features that may exist deep in the atmosphere but are not visible at the surface. One important set of observations confirmed that, unlike any other planet, Uranus rotates on its side. Astronomers had speculated that this was the case based on observations of the orbits of the moons of Uranus. The rotation axis is tilted just beyond 90 degrees relative to the planet's orbital plane. Instead of spinning nearly upright—the position predicted for every planet by the solar nebula model—Uranus rolls like a bowling ball, rotating east to west (retrograde) every seventeen hours for its sidereal day. As with

Source: NASA

Figure 7.8 Uranus and rings.

Source: NASA

Figure 7.9 Rings of Uranus.

the other gas giants, Uranus rotates faster at certain latitudes. A full rotation takes just over seventeen hours at the equator, while at about 60 degrees south, a full rotation appears to take as little as fourteen hours. We see this differential rotation with objects that have fluid surfaces, such as the gas giants and stars.

This rolling motion has a number of interesting effects. If a "day" on Uranus is just the time for one rotation, then the day is seventeen hours. However, that is the sidereal day. A solar day is from sunrise to sunrise or noon to noon or whatever other complete cycle of light and darkness you want to choose. On Uranus, that is the full eighty-four years of its revolution period! With its axial tilt, either its north or south pole is pointed almost directly at the Sun at different times in its orbital period. When one pole is going through "summer" on Uranus, it will experience forty-two years of continuous sunlight. When that same pole is pointed away from the Sun ("winter"), it will experience forty-two years of continuous darkness. Consequently, Earth-based telescopes are able to look almost directly down the polar axis of Uranus every forty-two years during the planet's summer or winter solstice. Since the faint rings orbit around Uranus' equator, they too are sideways, as are most of the planet's satellites.

This may come as no surprise, but given that a sideways planet does not fit the model, astronomers advanced a theoretical solution in the form of another large impact hypothesis. Their hypothesis suggests that Uranus formed upright on its axis, as the evolutionary perspective predicts, but then an asteroid the size of planet Earth crashed into it and knocked it over. The problem is that there is a general lack of evidence to support such an impact, and plenty of evidence to contradict the idea. First, Uranus is quite stable as it rolls along through space. Its orbit around the Sun is almost perfectly circular. Not only that, but its orbit lies more closely within the plane of the ecliptic than any other planet besides Earth. Massive collisions don't produce such perfect orbits. Second, Uranus has twenty-seven moons, and a faint ring system, orbiting nicely at Uranus's equator. Regardless of whether we try to weave a tale of those bodies joining Uranus before, during, or after the collision, we quickly hit logical and observational barriers. There is no evidence of a catastrophic collision. As a Nobel prize winning physicist said, "I think we should recognize that no collision occurred. Uranus was apparently created just the way it is."

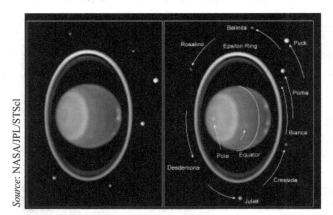

Source: NASA/JPL/STScI

Figure 7.10 A *Hubble* infrared image showing the "sideways" rings and satellite orbits.

Prior to *Voyager 2*, astronomers had predicted that Uranus would not have a magnetic field. After all, if Uranus has no internal heat source then there is no power source for the internal dynamo they say is needed to generate a magnetic field. However, *Voyager 2* showed that somehow this old, dead planet still had a magnetic field. This became all the more baffling when they discovered that the magnetic field axis of Uranus is offset from the rotation axis by an astonishing 60 degrees. The dynamo models predict that the magnetic field axis must be fairly well aligned with the rotation axis, so Uranus violates this condition as well. Secular theorists scrambled to provide an alternative explanation, suggesting that the dynamo effect might be occurring near the surface of the mantle rather than at the core. On the other hand, the magnetic field of Uranus does fit with biblical creation. In 1984—before *Voyager 2*, creation physicist Russell Humphreys predicted the magnetic field of Uranus based on the amount of magnetic decay that would have happened on the planet over a period of 6,000 years. *Voyager 2* confirmed this prediction. Although the presence of a strong magnetic field on any planet suggests a recent creation, this is especially true for Uranus.

Moons

Of the twenty-seven known moons, the two biggest and brightest moons of Uranus were discovered by William Herschel on a bitterly cold January night in 1787. While Herschel did not name them, his son John later would, settling on Oberon and Titania. You might recognize the literary

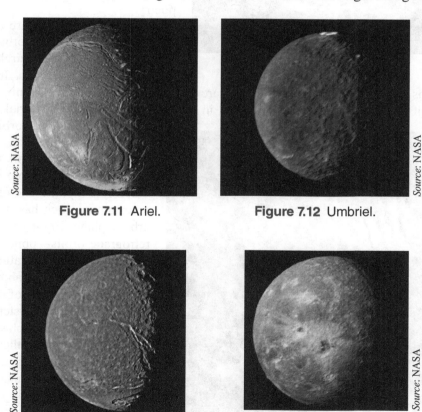

Source: NASA

Figure 7.11 Ariel.

Source: NASA

Figure 7.12 Umbriel.

Source: NASA

Figure 7.13 Titania.

Source: NASA

Figure 7.14 Oberon.

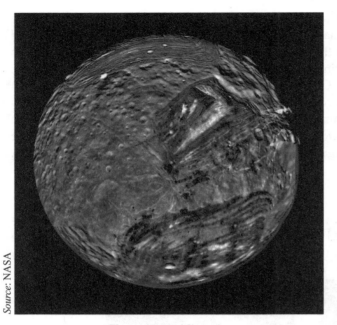

Source: NASA

Figure 7.15 Miranda.

roots from Shakespeare. Although they are the largest, they are less than half the diameter of Earth's Moon. Ariel and Umbriel were discovered in 1851, and strange little Miranda was discovered in 1948. These are the five major moons of Uranus, with the remaining twenty-two moons being much smaller, all less than about 100 miles in diameter and generally having irregular shapes. They were discovered during or after the *Voyager 2* flyby in 1986. The naming of these moons began a new tradition—they are all named after Shakespearian characters (including Juliet and Puck) or characters from Alexander Pope's poetry.

The Uranian moons are all composed of rock and ice. Thirteen orbit close to the planet in nearly perfect circles. Next out are the five major moons. In order from nearest Uranus and then moving outward, they are Miranda, Ariel, Umbriel, Titania, and Oberon. They too orbit in nearly perfect circles and all in the planet's sideways equatorial plane. These eighteen inner moons orbit prograde—in the same direction that Uranus rotates. Beyond Oberon, there is a relatively large gap before we reach the other nine moons. The orbits of these outer moons each has its own unique orbital plane. Eight of the nine have retrograde orbits—opposite the direction of the planet's rotation.

Miranda is only about 300 miles across but it's one of the strangest objects in the solar system. The entire moon looks like a patchwork of different terrain, glued together from a jumble of different pieces. Some sections look like someone painted them with a giant paintbrush, while other sections have very dramatic terrain like the ice cliffs with canyons several miles deep. As one evolutionist said, "no one

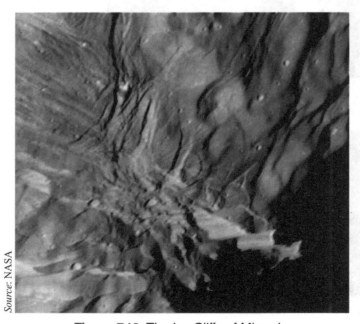

Source: NASA

Figure 7.16 The Ice Cliffs of Miranda.

predicted anything looking like Miranda." Another said, "the central problem in modeling the thermal histories of the Uranian satellites is accounting for Miranda." You might suspect that a secular explanation for Miranda involves an asteroid collision, but Miranda is so perplexing that some evolutionists propose *five* collisions! Here's a quote from a NASA website: "Scientists believe that Miranda may have been shattered as many as five times during its evolution. After each shattering the Moon would have reassembled from the remains of its former self with portions of the core exposed and portions of the surface buried." Creating an explanation by simply adding more collisions is not at all unlike Ptolemy adding more epicycles. It may be acceptable within the paradigm, but that doesn't make it realistic or convincing.

7.2 Neptune

If discovering Uranus was one of the great joys of William Herschel's life, working out the specifics of its orbital path and behavior was one of the great burdens. Herschel and other astronomers were satisfied with the general orbital features of Uranus when it came to identifying it as a planet. However, they soon found that further orbital calculations were not matching their observa-

Figure 7.17 *Voyager 2* provided these images of the "ice giants," Uranus (left) and Neptune (right).

tions. Using old observations from when Uranus had been thought to be a star, combined with new, ongoing observations, Herschel was perplexed. The observations did not match the predictions. What he saw happening was not matching up with the demands of Newton's laws. It was proving to be impossible to predict the future positions of Uranus with the degree of certainty that was expected. Shortly before William died, it was suggested to him that perhaps a large comet had struck Uranus just prior to its discovery, and the impact had altered its course. Nevertheless, William died in 1822 without enjoying any resolution to the problem that plagued his final years. His younger sister Caroline, on the other hand, would live to see things set right. The problem with Uranus was the presence of Neptune.

Discovering a Planet on Paper

The fanciful suggestion of a collision led to new calculations following Herschel's death that were a complete failure. So while the astronomers who were patiently making observations night after night had reached a crisis point with Uranus, theoreticians attacked the problem with pencil and paper. What if there was some other object—not yet discovered—whose gravitational influence could explain the peculiarities of Uranus' orbit? We call this influence of one object on the path of another a *perturbation*. Two men in particular, a Brit in his twenties (John

Couch Adams) and a Frenchman in his thirties (Urbain Jean-Joseph LeVerrier) independently worked on the problem, enduring laborious calculations, but in the end reaching solutions.

Once this new object—the object perturbing the path of Uranus—was located mathematically on paper, its existence needed to be confirmed in the night sky, and this is where Adams and LeVerrier ran into more difficulty. Then, as now, not just anyone could get time on a research telescope. In the case of Adams, he was an inexperienced newcomer, and when he approached the head of England's Royal Observatory at Greenwich he was initially ignored. LeVerrier on the other hand was known in Parisian science circles, and he had published his calculations, but he too was ignored by his national observatory. While Adams could do little more than wait, LeVerrier pursued another option.

If there is any truth to the saying, "it's who you know," it is borne out here. In 1835, LeVerrier had reviewed the thesis of a German graduate student named Johann Galle, and they became friends. Now Galle was an observer at the German national observatory in Berlin, so LeVerrier hoped for a favor. Galle was not the director of the observatory, so there was another hurdle to clear. The story goes that the director was rushing to head home for a birthday celebration in September of 1846 when Galle asked for permission, and perhaps because it seemed the most expedient, director Encke approved the request and left. Using LeVerrier's calculations, Galle located Neptune in less than an hour and began tracking its movements.

So who discovered Neptune? History suggests that Adams successfully completed his calculations first, but his national observatory dragged their feet and started work too late. LeVerrier had the accurate calculations, but what good were they without a telescope? Galle was the first person to see Neptune as a planet, but would he have ever seen it without LeVerrier's calculations? We believe that Galileo saw Neptune in the 1600s, but saw it as a star, just as Uranus had been misidentified for so long. So Galileo saw it first, but didn't know it was a planet. While other sources may not agree, the proposal here is that all three men—Adams, LeVerrier, and Galle—deserve a share of the credit for the discovery of Neptune. Unlike Uranus, which was a "natural" (accidental?) discovery, Neptune required the calculations of theoreticians as well as telescopic confirmation. Perhaps Sir Isaac Newton deserves a share of the credit as well, since his laws made the calculations possible.

After 1846, credit for the discovery became a point of contention between England and France, although Adams and LeVerrier harbored no such feelings. The French boasted of their discovery, conveniently overlooking their refusal to grant access to their telescope and forcing LeVerrier to approach Galle. England pointed to the calculations Adams had completed ahead of LeVerrier, as if that somehow compensated for the inept behavior of the director of their own national observatory. Perhaps it is also fitting that once again there were arguments about the name. Neptune, Oceanus, and LeVerrier were all in the running, but of course Neptune, the god of the sea, won out in the end.

The Second Ice Giant

The discovery of Neptune stretched our solar system out to 30 AU. At this distance it takes light from the Sun approximately four hours to reach the planet. As Kepler's third law of planetary motion predicts, it takes 165 Earth years for Neptune to make one trip around the Sun. It has traveled just over one revolution since its discovery in 1846. A sidereal day on Neptune is about sixteen hours. Like all the gas giants we find an atmosphere of hydrogen, helium and other trace elements, multiple moons and a ring system. **It's just slightly smaller, but more massive, than Uranus, and it has a pretty blue shading, reminiscent of watery oceans. But the blue is not water, just methane gas, reflecting blue light back to our eyes.** Like Uranus, we can also refer to Neptune as an ice giant.

Figure 7.18 Neptune and its rings.

Neptune possesses an incredibly stormy atmosphere. When *Voyager 2* encountered Neptune in 1989, it found an Earth-sized hurricane blowing in the southern hemisphere. It was called, simply, the **Great Dark Spot. It was assumed that the Great Dark Spot was a fairly permanent feature in the Neptunian atmosphere, like the Great Red Spot found on Jupiter. But when the *Hubble Space Telescope* looked for it in the mid-1990s, it had disappeared! Meanwhile, new storms and another dark spot had formed in Neptune's northern hemisphere. In other areas, great cirrus clouds stretch for thousands of miles across the planet's sky. Neptune has the fastest wind speeds yet recorded: 1,500 miles an hour!** More baffling than the wind speeds themselves is that these winds are going in the opposite direction of the planet's rotation. Why Neptune has such high-speed retrograde winds is not understood at this time.

One of the first discoveries of *Voyager 2* was the detection of a ring system. The existence of Neptune's rings was not a complete surprise, but *Voyager 2* was able to confirm and directly image them. At a great enough distance, these rings looked more like arcs rather than complete circles. *Voyager 2* showed that the rings are complete, but they vary in thickness. It turns out that Neptune has five major rings that are a mosaic of the types of rings encircling the other gas giants. Three of them are thin threads, as we see around Uranus. The other two rings are in

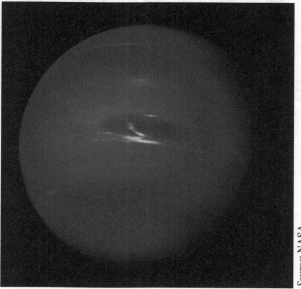

Figure 7.19 Neptune's Great Dark Spot.

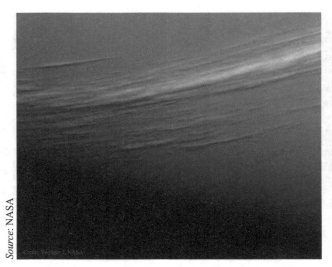

Figure 7.20 Cirrus clouds.

Figure 7.21 Neptune's rings.

broad sheets, as we saw with Saturn's rings, but they are very thin like Jupiter's. The rings are named for people who were involved in some way with the planet's discovery: Galle, LeVerrier, and Adams, who have been discussed at length, Lassell, who discovered Neptune's largest moon, and oddly enough, Arago, who failed to give LeVerrier access to the telescope in Paris but then later wanted the planet to be named LeVerrier.

Recall that Uranus was the one gas giant that did not radiate more energy than it was receiving from the Sun. While Neptune is basically a clone in terms of size and composition, it is radiating more than twice the energy it receives from the Sun, which means it has considerably more internal heat. As was the case with Jupiter and Saturn, it is hard to imagine how this process could last for billions of years, but is not a problem for the biblical timescale.

Voyager 2 measured the magnetic field of Neptune and found it to be similar in strength to what it had measured for Uranus. This is consistent with each planet's biblical age of thousands of years but is far stronger than what we would expect if the planets were billions of years old since magnetic fields decay over time. As with its twin Uranus, Neptune's magnetic field is also tilted and offset from the planet center. This is extremely difficult to account for in secular dynamo models. When *Voyager 2* found that the magnetic field of Uranus was tilted and offset from the center of the planet, evolutionists said maybe *Voyager* flew by just as Uranus's magnetic field was reversing. This would have been very unlikely, although not impossible. However, three years later *Voyager* arrived at Neptune and saw the same thing. For all practical purposes, the likelihood of observing two planets that are both experiencing magnetic polarity reversals is zero.

Moons

Given Neptune's distance, only two moons were confirmed with ground telescopes before *Voyager 2*. The first and largest one, Triton, was actually discovered only seventeen days after the planet itself was found, thanks to the work of British astronomer William Lassell. This provided all that astronomers needed to calculate the mass of Neptune. *Voyager 2* then discovered five new moons orbiting close to Neptune. These are all small, at less than 300 miles in diameter. With technological breakthroughs in ground-based imaging, several additional moons were discovered in the years following the *Voyager 2* encounter, bringing the number of known moons for Neptune to fourteen.

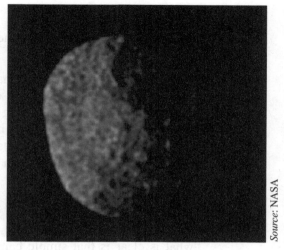

Source: NASA

Figure 7.22 Neptune's moon Proteus was discovered by Voyager 2.

Triton is about one-fourth the diameter of the Earth's Moon, making it the seventh-largest moon in the solar system. Unlike any other large moon, Triton's orbit is retrograde—opposite the direction that the planet spins. Also, large moons tend to orbit in the plane of their planet's equator (zero degrees), but Triton orbits at an angle of 23 degrees. Some astronomers have suggested that these idiosyncrasies are the result of a collision. (That cannot possibly come as a surprise by now.) Only Triton's size made it possible to be discovered in the nineteenth century. It would be over a century before a second moon could be detected.

Source: NASA

Figure 7.23 Triton.

In some areas on Triton's surface there is a mottled, "cantaloupe terrain" formed by rising blobs of ice. There are also volcanoes on Triton that are unlike anything else. They don't erupt fire and molten lava, but jets of liquid nitrogen! These structures have come to be known as **cryovolcanoes ("cryo" means, "cold). Dark, streaky carbonaceous deposits can be found downwind from the craters, on what is surely the coldest surface ever recorded: nearly 400 degrees below zero!** This sort of activity is quite surprising if the moon is billions of years old.

Is Neptune Really There?

According to evolutionary models, Neptune should not exist. Here's how *Astronomy* magazine explained it:

> "Astronomers who model the formation of a solar system have kept a dirty little secret: Uranus and Neptune don't exist, or at least computer simulations have never explained how planets as big as the two gas giants could form so far from the Sun. Bodies orbited so slowly in the outer parts of the Sun's protoplanetary disk that the slow process of gravitational accretion would need more time than the age of the solar system to form bodies with 14.5 and 17.1 times the mass of Earth."

Evolutionists are of course disappointed that their model fails. As one evolutionary astronomer has commented,

> "what is clear is that simple banging together of planetesimals to construct planets takes too long in this remote outer part of the solar system. The time needed exceeds the age of the solar system. We see Uranus and Neptune but the modest requirement that these planets exist has not been met by this model."

This problem was first discovered by a scientist named Safronov, who published his findings in 1972. His work suggested that the time for these planets to form based on the solar nebula model was easily double the proposed age of the solar system. Depending on your perspective, 1972 was a long time ago, and certainly enough time for astronomers to come clean about the accepted model of how solar systems form. The problem isn't that the model fails. That is just part of science. The problem is pretending that the model doesn't fail.

Here's a quote from another evolutionary astronomer:

> "It's clear that our level of sophistication of studying planet formation is relatively primitive. So far it's been very difficult for anybody to come up with a scenario that actually produces Uranus and Neptune."

That is an interesting phrase: "come up with a scenario." Rarely do you find communication with the public from the scientific community that suggests that the solar nebula model, the big bang theory, or prebiotic evolution are anything less than proven facts. That's why it is supposed to be appropriate to ridicule anyone who disagrees. After all, we must "follow the science." "Science is truth," right? The reality is that often what these models are attempting to do is "come up with a scenario" that sounds plausible if you have embraced naturalism. As we saw in an earlier chapter, historical science is by its very nature highly speculative and strongly influenced by a priori philosophical commitments. In this case, a commitment to naturalism makes someone willing to accept a model that predicts that Uranus and Neptune should not exist, if the alternative is a Creator.

These planets reveal something about the creation versus evolution debate. This is not a debate about religion versus science. We have a long history of great scientists who were also creationists. There are creationists today in every field of science. Ultimately, the debate is about the authority of the Bible as God's Word. As one creationist noted, "there is no lack of adequate proof of our Creator's existence, power, wisdom, care, and glory. Rather, those who choose to be unpersuaded are, as the apostle Paul says, 'without excuse.'"

7.3 What about Pluto?

When independently wealthy Percival Lowell built his own research observatory in Flagstaff, AZ, it was with the immediate plan of observing the planet Mars as it approached its 1894 opposition. Lowell had become obsessed with all things Mars, and was convinced that it harbored intelligent life. Once the observatory was complete, he spent a year making observations and drawing illustrations for the book he would then publish, titled simply *Mars*. Two other books about the red planet would soon follow. Of course, most of what Lowell wrote lacked any true merit, and exposed him and his observatory staff to ridicule. Lowell seemed immune to such things, but his staff was demoralized by 1901, when Lowell decided to change directions.

Figure 7.24 Percival Lowell.

Harvard educated, Lowell excelled at mathematics. His new project would be to replicate what Adams and LeVerrier had done years earlier. He would do the calculations necessary to find a ninth planet, which Lowell called "Planet X." To be fair, at this time astronomers were convinced that there was another planet influencing Uranus and Neptune, so why not? Lowell had the drive, the staff, the wealth and the first-rate observatory, and this search was completely credible in the astronomical community. Lowell did the calculations and faithfully searched—unsuccessfully—until his death in 1916. Even in death,

Figure 7.25 The Lowell Observatory in Flagstaff, AZ.

Lowell made arrangements to continue funding the observatory and the research using his sizeable fortune. Although his widow held things up for a decade contesting the will, by 1929 the observatory had a new telescope and a new astronomer to continue the search.

Clyde Tombaugh was a farm boy from Kansas, and by all accounts a genuinely decent person. He responded to the search from the Lowell Observatory with materials he had generated as an amateur astronomer, and although he had no degree and no professional experience, he was hired. Tombaugh used what money he had to buy a train ticket to Flagstaff and begin difficult work at a low paying job. LeVerrier's calculations allowed Galle to find Neptune in less than an hour. Lowell's calculations proved to be less forgiving. For ten months, Tombaugh spent his nights in the cold air, taking hour-long exposures of the sky. What would follow then was the tedious process of looking at the developed plates using a machine called a *blink comparator*. This device allows a view of two photographic plates in rapid succession. By shifting from one photograph to a nearly identical one taken some time later, Tombaugh would be able to perceive any change. A planet will appear to jump back and forth when the plates are flipped, while the stars remain stationary. In 1930, his patience and persistence paid off. He found Planet X, about one billion miles beyond Neptune (40 AU), with an orbital period of 248 years. The twenty-four-year-old Tombaugh made history, and was then given a leave to earn his degree in astronomy at Kansas University.

Image courtesy of NASA

Figure 7.26 Clyde Tombaugh.

Courtesy NASA

Figure 7.27 In this size comparison you can see the difference in size between Pluto, Earth, and the Moon.

In the years that followed, several things came to light. First, Pluto was not what anyone expected. It was incredibly small compared to the gas giants, and as observations improved, new calculations only made it smaller. It had a highly eccentric orbit (coming closer to the Sun than Neptune when at perihelion) and a very inclined orbital path (it will never crash into Neptune). At best it was a very odd planet. Second, Lowell's calculations were never correct from the start. By the time *Voyager 2* passed Neptune in 1989, astronomers could see that their calculations had been off, and there was no problem with the orbits of Uranus and Neptune. Lowell's calculations were based on a false premise, and had no more value than his book about Mars. This makes Tombaugh's discovery all the more remarkable. Third, Pluto was not alone.

Starting in the 1990s, astronomers started finding more objects in the same "neighborhood." It is this final observation—that Pluto was not alone—that would lead to its demise.

In 2006, astronomers officially decided that Pluto was not a planet. What happened, and why did astronomers make this change? When compared to the terrestrial planets and the gas giants, Pluto is all over the place with its characteristics. This is one reason why astronomers were suspicious of Pluto—it did not fit in with the classification scheme characteristics of the planets. Another concern which was already mentioned is that Pluto is quite small. Earth has 400 times more mass than Pluto, and Mercury, currently considered the smallest planet, is 22 times more massive than Pluto. For sixty years after its discovery in 1930, Pluto remained the sole known object with a planet-like orbit orbiting the Sun beyond Neptune. But in the early 1990s astronomers began finding many other such objects. Many of those objects have sizes comparable to Pluto, and in 2003 astronomers found one object, "Eris," that was slightly larger than Pluto. If Pluto was a planet, then there is no good reason why these other large objects are not planets too. This could quickly lead to there being scores and even hundreds of planets. Astronomers thought that they ought to draw the line at some point. In 2006, the International Astronomical Union voted to define what a planet is in a manner that excluded Pluto. This was a controversial decision, and astronomers may be revisiting and updating this issue in years to come.

So What is Pluto?

*If Pluto is not a planet, then what is it? With the discovery of so many objects beyond the orbit of Neptune, it appears that this is another asteroid belt (the one between Mars and Jupiter has been known since the nineteenth century), and Pluto is just one of the larger members of this belt. There are different classifications for these bodies. Some astronomers think that they are the **Kuiper belt**, the supposed source of short period comets, so they call them KBOs (Kuiper Belt Objects). Other astronomers prefer TNOs (**Trans Neptunian Objects**). Pluto and a few of the other larger minor planets are spherical. This is unusual, because most minor planets are odd shaped. What causes a body to be spherical? A sphere is the minimum shape that a body can have if the body has sufficient gravity. To have sufficient gravity, a body must have enough mass. Most minor planets lack the mass to produce sufficient gravity to make them spherical. Astronomers coined a term to describe spherical minor planets such as Pluto—**dwarf planets**.

Despite common misconception, the demotion of a planet is not new. The first minor planet, Ceres, was discovered in 1801 in between Mars and Jupiter, a location where many people expected a planet to be. Astronomers called Ceres a planet for more than forty years, even after a few more asteroid belt objects were discovered. By 1845 astronomers concluded that, much like Pluto, Ceres had too many "neighbors" to be a planet. Incidentally, since Ceres is spherical, it too is a dwarf planet. The number of dwarf planets is sure to increase in number over the years.*

New Horizons

In 2015, when the *New Horizons* spacecraft completed its nine-year journey to Pluto, what the NASA scientists saw was stunning. Prior to this, we had poor-quality images, even using *Hubble*.

Figure 7.28 The *New Horizons* mission produced images that were far better than anything before. Here we see Pluto with an image of its largest moon, Charon in the background.

Image courtesy of NASA

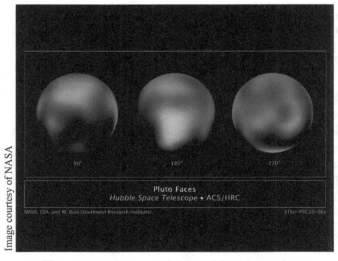

Image courtesy of NASA

Figure 7.29 By comparison, this was our best imaging prior to the *New Horizons* pictures.

We knew Pluto had one relatively large moon (Charon) and several quirky, smaller moons. Beyond that, the expectations were pretty straight forward—Pluto would be an old (4.6 billion years), cold, geologically inactive object. While the average surface temperature is just minus 380°F, "cold" here refers to the planet's internal heat. Remember, the solar nebula model says smaller objects lose their heat faster, so an "old" Pluto should have no internal heat to drive any geological activity. *New Horizons* destroyed all of the expectations.

When *New Horizons* began sending back photos, secular scientists were shocked. Pluto is a dynamic, geologically active place. Large sections of Pluto's surface are smooth and craterless. These regions are very young because there hasn't been enough time for craters to form in them. As one NASA project scientist said, "this is one of the youngest surfaces we've ever seen in the solar system." Recent studies have calculated that their *maximum* age is 500 thousand years. In reality, it could be much younger than that. In fact, as one scientist admitted, "this could be only a week old for all we know."

Apparently, Pluto is resurfacing itself. It appears that this is happening by convection, with warm material rising up from within. This was a huge surprise as Allen Stern—the New Horizons principal investigator—said, "Sputnik Planum [the region where the convection is happening] is one of the most amazing geological discoveries in 50-plus years of planetary exploration." *National Geographic* commented, "it's a geologically rapid process that scientists didn't exactly expect to see on a small freezing world that lives on average 40 times farther from the Sun than Earth."

Convection requires internal heat, but where such heat would come from is a mystery. How can Pluto still be geologically active after billions of years? As Alan Stern said, "finding that Pluto is geologically active after four-and-a-half billion years . . . there's not big enough typeface to write that in . . . it's unbelievable." The only reasonable explanation for Pluto's geological activity is primordial heat left over from when it was formed, but that can only last for a relatively short time, not 4.6 billion years.

Figure 7.30 Some of the "young" icy mountains on Pluto.

Image courtesy of NASA

Pluto also has "young" mountain ranges but these mountains are a mystery to scientists who believe in billions of years. These mountains appear to have been formed recently. As one article noted,

> "the prominent mountains at the western margin of Sputnik Planitia and the strange multi-kilometer-high mound features to the south are both young geologically and presumably composed of relatively strong water ice-based geological materials. Their origin and what drove their formation so late in solar system history remain uncertain."

Pluto has cryovolcanoes, actively flowing glaciers, windblown dunes, and unusual features called "snakeskin (or 'bladed') terrain." None of these could have been expected for an "old" object on the outer edge of our solar system. There were enough varied, unusual, and spectacular geological features discovered on the dwarf planet, that NASA gave this succinct summary in 2020: "In the five years since that groundbreaking flyby, nearly every conjecture about Pluto possibly being an inert ball of ice has been thrown out the window or flipped on its head."

In 1978 we discovered that Pluto had a relatively large moon, Charon. Charon is the largest moon when compared to the size of its companion. In fact, some astronomers consider the two to be a binary system of dwarf planets. A common explanation for its origin is—of course—a catastrophic collision early in Pluto's history. *New Horizons* revealed

Figure 7.31 This "bladed terrain" had never been seen anywhere but Earth prior to these images from Pluto.

Image courtesy of NASA

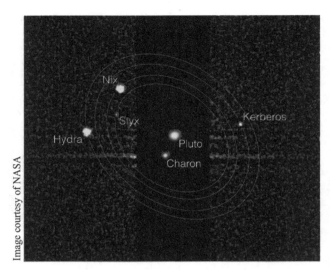

Image courtesy of NASA

Figure 7.32 Pluto has five satellites.

some of the same problems regarding Charon for the secular models that Pluto does. Much of Charon's surface looks young and there appears to be recent geological activity. As a NASA article said,

"Mission scientists are surprised by the apparent lack of craters on Charon. Relatively few craters are visible, indicating a relatively young surface that has been reshaped by geologic activity . . . that Charon is so geologically complex, however, would seem to require a heat source or reshaping what would have otherwise been the heavily cratered surface."

Here is the same problem: if Pluto should have cooled off many years ago, Charon should have too, yet it still contains leftover heat from its creation. If that was billions of years ago this is a problem but if that creation took place just thousands of years ago, then it's not surprising.

Pluto has four other smaller moons: Styx, Nix, Kerberos, and Hydra. These moons look young. They are very bright, especially Nix and Hydra, but this brightness wouldn't last for long. As one article said,

> "how such bright surfaces can be maintained on Nix and Hydra over billions of years is puzzling given that a variety of external processes (e.g. radiation darkening, transfer of darker material from Charon via impacts, impacts with dark Kuiper belt meteorites etc.) would each tend to darken and redden the surfaces of these satellites over time."

Pluto is a wonderful example of how a flawed model leads to flawed predictions. Nearly everything scientists predicted and expected about Pluto was proven to be incorrect by the *New Horizons* spacecraft. It is not hyperbole to say that as the first gorgeous images of Pluto were coming in, NASA scientists were shocked. You could see it on their faces, and you could hear it in their voices. They admitted their surprise, but even now they continue to try to make the new observations fit with the old model. You can never fit a creation-shaped event into a naturalism-shaped hole.

7.4 Comets, Asteroids and Meteors
Comets

*Comets are small icy bodies with an admixture of very tiny solid dust particles. Their densities are very low, suggesting that they are very porous and fragile. Astronomers call this the "**dirty iceberg" theory**. Actually, what we have described is the nucleus of a comet (Figure 7.33). As we previously mentioned, comet orbits are very long ellipses. The Sun is near one end of the ellipse. The point on the orbit nearest the Sun is the perihelion, and the point farthest from the Sun is aphelion. A comet moves very quickly when near perihelion, but it moves very slowly when near aphelion. As a result, comets spend most of the time very far from the Sun. When so far from the Sun, the material in a comet nucleus remains frozen. However, once each orbit when a comet nucleus approaches perihelion, the heat of the Sun evaporates much of the ice directly into gas without going into a liquid phase (we call this sublimation). Much dust dislodges along with the gas. The gas expands into a large cloud around the nucleus, and excitation by solar radiation causes the cloud to glow. We call this cloud the **coma**. A comet's coma may be many tens of thousands of kilometers across, even though the

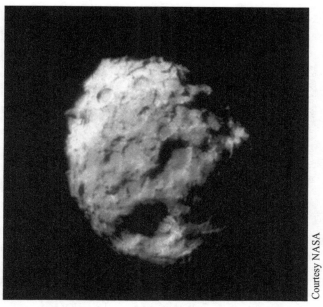

Figure 7.33 The nucleus of Comet Churyumov–Gerasimenko as imaged by the Rosetta mission.

Courtesy NASA

nucleus is only a few kilometers across. Solar radiation pushes the microscopic dust particles away from the Sun, producing a curved **dust tail**. The solar wind, an out-rush of charged particles from the Sun, pushes the gas away from the Sun to produce a gas, or **ion tail**. The dust tail shines by reflecting sunlight, while the ion tail glows from ionization from the Sun's radiation (Figure 7.34).*

*The greatest brightness of a comet typically occurs shortly after it passes perihelion. However, how bright a comet appears to us on Earth depends upon its distance from Earth and how far from the Sun it appears in our sky. On rare occasions, a comet can appear extremely bright, and its tail can extend far across the sky. However, a comet looks best in a dark sky, which can be difficult for many city dwellers to achieve. Until recent times, people viewed comets as bad omens that portended disasters. This likely resulted from what appeared to be very erratic behavior of comets. The Sun, Moon, and planets follow complicated motions, but they are confined to near the ecliptic and are straightforward to understand. However, comets tend to appear without warning, move across the sky in odd directions, and then disappear never to be seen again. Of course, comets

Figure 7.34 Comet Hale-Bopp is sometimes called the Great Comet of 1997.

Valerio Pardi/Shutterstock.com

Image courtesy of NASA

Figure 7.35 You may have seen Comet NEOWISE for yourself in 2020.

follow the same sort of physical laws that the planets do. However, comets have very elliptical orbits that are highly inclined to the ecliptic, and they have longer orbital periods than the naked-eye planets. These factors result in any particular comet being visible for only a short portion of each orbit near perihelion, making its motion appear erratic.

Comet nuclei are very small (only a few kilometers across) and hence contain very little mass. Yet, a comet loses a significant portion of its mass each pass near perihelion. It is obvious that a comet cannot orbit the Sun many times before losing so much of its volatile material (the ices that sublime and then fluoresce to produce the bright coma and ion tail) that the comet fails to shine anymore. Indeed, some comets with periods short enough to have been observed on several returns to perihelion have grown progressively fainter over the centuries. The most famous comet, Halley's, with a period of seventy-five to seventy-six years, is a good example of this gradual fading. A conservative estimate of how many trips that a comet can make around the Sun and still be visible is about one hundred. There is a maximum orbital size for a comet. If a comet's aphelion is too far from the Sun, the gravitational perturbations of other stars can permanently remove a comet from the Sun's grasp. The maximum orbital size to prevent this from happening corresponds to an orbital period of a few million years. Multiplying these two figures, we find that comets cannot be visible after they have been orbiting the Sun for more than a few hundred million years. Since most scientists think (based upon various old-Earth assumptions and methods) that the solar system is 4.6 billion years old, there ought not to be any comets left.*

*The situation is even worse for an old solar system, for there are two other mechanisms whereby comets are lost to the solar system. The gravitational perturbations of Jupiter and the other Jovian planets frequently add energy to a comets orbit, effectively ejecting them from the solar system, never to return. Astronomers have observed ejection of comets many times, and this represents a catastrophic loss of comets. Gravitational perturbations can rob a comet of orbital energy, which results in a shorter period. This results in more frequent trips to perihelion, with even faster gradual wearing out. Another catastrophic loss of comets is collisions with planets. Astronomers have observed this once. In 1994, Comet Shoemaker-Levi IX slammed into Jupiter. Two years earlier, this comet passed close to Jupiter on its way toward perihelion. Tidal forces of Jupiter shredded the comet into more than two dozen pieces. As those pieces moved away from perihelion, they struck Jupiter over several days, leaving behind dark marks that persisted for several days. This comet no longer exists.

Evolutionary astronomers are well aware of this problem, so they have offered two solutions to it. About 1950, the Dutch astronomer Jan Oort suggested that the Sun is surrounded by a spherical cloud of comet nuclei, with the nuclei orbiting far from the Sun (thousands of AU). Today we call this hypothetical distribution of comet nuclei the **Oort cloud**. The gravitational perturbations of passing stars or other objects in the galaxy were supposed to alter the orbits of these nuclei so that they fell to the inner solar system to become visible as comets. Today astronomers think that the dominant mechanism of injecting comets into the inner solar system is tides produced by the galaxy. As old comets died out, new ones would rain down to establish something close to a steady state. Oort suggested his cloud to explain the origin of long period comets, comets generally with periods greater than about 200 years. This is not an arbitrary distinction, for long period comets generally have orbits highly inclined to the ecliptic, with about half of them orbiting the Sun the same direction that planets do, but half in retrograde orbits.

Short period comets, comets with periods less than about 200 years, tend to have low inclination orbits and orbit prograde, the same direction that planets orbit the Sun. About the time that Oort proposed his cloud, the Dutch-born American astronomer suggested a belt that bears his name for the source of short period comets. The Kuiper belt supposedly lies just beyond the orbit of Neptune (recall that this is where Pluto is, making Pluto a member of KBOs in the estimation of many astronomers). Gravitational perturbations of the Jovian planets supposedly convert KBOs into short period comets. For about thirty years astronomers assumed that gravitational perturbations on long period comets could convert them to short period comets, so most of them assumed that the Kuiper belt was not necessary. However, by the early 1980s, astronomers had run simulations that showed that gravitational perturbations were far too inefficient to convert a significant number of long period comets into short period comets before they would be destroyed. Hence, astronomers began to accept the Kuiper belt and began to look for objects there, which resulted in the discovery of the

new asteroid belt (TNOs). Today, most astronomers think that the Kuiper belt is the original source of comets, and that planetary perturbations not only produce short period comets, but that those same perturbations populate the Oort cloud from which come long period comets.*

Does either the Oort cloud or the Kuiper belt exist? Even if the Oort cloud exists, it is unlikely that we will ever observe it, and by 2010 secular researchers were acknowledging "the standard model can't produce anywhere near the number of comets we see [falling in from the Oort Cloud]." *Are the TNOs actually KBOs? There is still much room for debate on this. Many TNOs appear to be far too large to be comet nuclei. Furthermore, we know the densities of a few KBOs (Pluto is

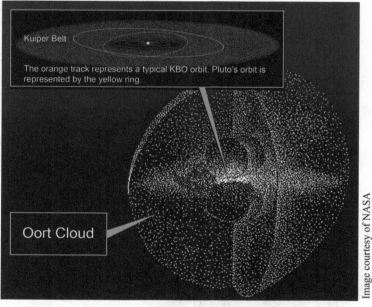

Figure 7.36 The lack of observational evidence does not prevent astronomers from "showing" us the Oort Cloud. Our solar system is tiny (at the center of the image) compared to the proposed size of the cloud.

Image courtesy of NASA

the best example). The inferred composition of TNOs does not match the composition of comets. At this time, it appears that the existence of comets still is a good argument for a recent origin of the solar system, suggesting that the solar system is far younger than billions of years.*

Minor Planets (Asteroids)

In 1801, **Father Giuseppi Piazzi of Sicily found the first asteroid telescopically and named it **Ceres**. Since William Herschel's discovery of Uranus in the late eighteenth century, astronomers had dusted off their telescopes and picked up the search for more objects in the solar system. One of the likeliest places to look was between the orbits of Mars and Jupiter, where a large, suspicious-looking gap existed. **Johannes Bode**, director of the Berlin Observatory, had popularized an interesting mathematical quirk first introduced by another astronomer named Titius. Start with the number 0. Then go to 3. Now double 3 to get 6. Keep doubling: 12, 24, 48, 96, and 192. Then add 4 to each of these numbers to get 4, 7, 10, 16, 28, 52, 100, and 196. Now divide those numbers by 10 to get 0.4, 0.7, 1.0, 1.6, 2.8, 5.2, 10.0, and 19.6. The planet Mercury is 0.4 AU from the Sun; Venus is 0.7 AU out; Earth is 1.0 AU; Mars is 1.5 AU; Jupiter is 5.2 AU, Saturn is 9.5 AU; and Uranus is 19.2 AU. Not a bad approximation, but there doesn't seem to be any planet at 2.8 AUs—that's the gap between Mars and Jupiter. Astronomers began to wonder, might there be another planet hanging out in that gap?

Piazzi wasn't looking for any new planets when he found Ceres; he was engaged in trying to locate a particular star that had been mis-catalogued. But there it was, an unlooked-for star that slowly changed its place among the stars over a successive series of nights. When the existence of Ceres was confirmed and its distance was determined to be 2.8 AUs, astronomers were ecstatic—**"Bode's Law"** worked (until Neptune really ruined things)! But it was quickly discovered that Ceres was not a particularly large object, only about 500 miles in diameter. And then more asteroids were discovered. William Herschel suggested the term asteroid ("*star-like*") for this new class of minor planets.**

*By the end of the nineteenth century, the number of minor planets had grown to about 300. By 1960, the number had grown to nearly 2,000. As of 2015, there were more than 600,000 known minor planets, and the discoveries continue unabated. Many of the ones now discovered are very small, for most of the larger ones were found long ago. The exceptions to this would be the ones among the TNOs. Most minor planets are in the two belts, one between Mars and Jupiter, and the other one beyond Neptune. However, there are many found orbiting elsewhere. Astronomers often classify minor planets according to their orbits. Of particular interest are the minor planets whose orbits cross Earth's orbit, for this could result in collisions with Earth. Astronomers call these **near-Earth asteroids**, or NEAs for short.*

Source: NASA

Figure 7.37 Ceres—the largest of the asteroids, also qualifies as a "dwarf planet."

*Astronomers also classify minor planets according to composition, as inferred from color, spectroscopy, or density measurements. Minor planets in the inner solar system, those orbiting closer to the Sun than the asteroid belt, tend to be made of denser material, suggesting composition similar to the terrestrial planets. The minor planets of the inner solar system further subdivide with regard to composition. Some primarily are made of iron and nickel. Astronomers term these M-type

Figure 7.38 False-color image of Occator crater and salt deposits on Ceres.

Source: NASA

minor planets, the M standing for metal. Other minor planets are rockier. They are S-type minor planets, the S standing for silicate, a major component of many rocks. Some rarer minor planets are C type, with C standing for carbonaceous. Many astronomers think that C-type minor planets may be burned out comets. Minor planets of the outer solar system, those beyond the asteroid belt, tend to be icier. This is similar to the composition of other small objects of the outer solar system, comets, and the satellites of the Jovian planets.* Some of these outer-planet objects are known as "centaurs." Just as the mythological centaurs were a mix of horse and human, the centaurs found crossing the paths of the outer solar system planets appear to have a mix of characteristics of comets and asteroids. They tend to be similar to asteroids in size, but more like comets in terms of their composition, adding to the difficulty of creating sharp distinctions between the categories.

**In addition to radar images of asteroids garnered from Earth-based observatories, the *Galileo* probe sent us close-up images of two asteroids—Gaspra and Ida. Ida was a rarity among these minor planets—it possessed a tiny moon of its own, called Dactyl. Many other asteroid moons have since been discovered.

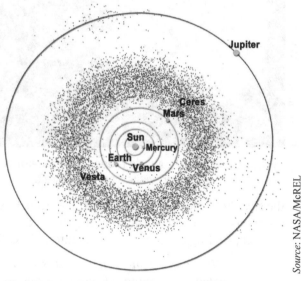

Source: NASA/McREL

Figure 7.39 Besides the main asteroid belt just beyond Mars, notice that some asteroids do occupy the inner part of the solar system. Also note the "Trojan Asteroids" sharing Jupiter's orbit roughly 60 degrees ahead and 60 degrees behind the gas giant. Not included here are the TNOs.

Courtesy NASA

Figure 7.40 The minor planet Ida has a small satellite, Dactyl.

Image courtesy of NASA

Figure 7.41 Vesta dwarfs these other asteroids that have been visited by spacecraft.

In 2001, the probe *NEAR-Shoemaker* soft-landed on the asteroid **Eros**, after having orbited this 20-mile-long rock for a full year. Although the probe was not intended for landing, Jet Propulsion Lab scientists were able to bring it down gently, thanks to the very low surface gravity of Eros—a 200-pound astronaut would weigh just 2 ounces on Eros, so light that he could just about launch himself into orbit by the power of his own legs.**

We also have detailed images of the third largest member of the asteroid belt, Vesta, thanks to the *Dawn* spacecraft that orbited this minor planet in 2011 and 2012. Its mission was later extended to include Ceres. A handful of other asteroids have also been visited by spacecraft, including Mathilde, Itokawa, and Lutetia. Spacecraft provide remarkable images of these tiny worlds that would not be possible with Earth-based telescopes.

Meteors

Folklore is filled with stories about people witnessing rocks falling out of the sky to Earth. But most scientists in the eighteenth century dismissed the notion as being ridiculous. Then in 1803, a singular event occurred. Thousands of meteors streaked across the sky above Normandy in France. The French Academy of Sciences sent Jean-Baptiste Biot and his assistants out to investigate. Biot interviewed eyewitnesses and from their accounts was able to map the area, finding hundreds of stones littering the field in an elliptical "footprint."

> **I saw the awful traces of this phenomenon: I traversed all the places . . . I collected and compared the accounts of the inhabitants; at last I found some of the stones themselves on the spot, and they exhibited to me the physical characters which admit of no doubt of the reality of their fall . . . In this account I have confined myself to a simple relation of facts . . . I have succeeded in placing beyond a doubt the most astonishing phenomenon ever observed by man.*

— **from Jean-Baptiste Biot's report to the French Academy of Sciences, 1803**

There was another meteor fall in the United States, during the Thomas Jefferson administration. This time ground zero was in the state of Connecticut. Two Yale professors investigated, and like Biot, they came to the conclusion that rocks had fallen to Earth from outer space. President Jefferson, trained in the sciences, allegedly said of the investigation, *"I would rather believe that two Yankee professors would lie than believe that stones fall from heaven."***

*Occasionally, a piece of a minor planet or a comet collides with Earth. Earth's atmosphere protects us from most of these. As these fragments enter the atmosphere at an altitude of about 100 km, they interact with molecules in the atmosphere to heat the air to a very high temperature. The heat is so intense that the air begins to glow around the quickly moving particle. We see the glow at night as a **meteor**, or shooting or falling star. The hot air erodes the incoming particle so that smaller particles completely burn up. If an incoming object is large enough, a portion may survive to the ground. If we find a surviving piece on the ground, we call it a **meteorite**. There are three basic types of meteorites: irons, stony, and stony-irons.*

An iron meteorite is a piece of an M-type minor planet. **Irons** are made typically of iron and nickel. They're easier to identify than stonys, as they really do have an out-of-this-world appearance. While only about one out of ten rocks that hit the Earth are irons, they're the ones usually picked up. Often when you polish and etch iron meteorites with a mild acid solution, you'll get something called a Widmanstätten pattern, named for an artisan who discovered the process, although William Thomson happened

Figure 7.42 Meteoroids become meteors when they burn up in the Earth's atmosphere.

© Vadim Sadovski/Shutterstock.com

Image courtesy of NASA

Figure 7.43 A meteor results when a fast-moving small particle of matter burns up high in Earth's atmosphere.

© Linnas/Shutterstock.com

Figure 7.44 Iron meteorite.

© Albert Russ/Shutterstock.com

Figure 7.45 Widmanstätten pattern in polished iron.

© Linnas/Shutterstock.com

Figure 7.46 Stony meteorite.

onto the same process four years earlier, in 1804. These are iron crystals arranged in a kind of cross-hatched pattern.**

A stony is a part of an S-type minor planet. **Stony** meteorites tend to look a lot like rocks you'd find in your driveway, so they are not often recognized as having come from outer space. They are not as durable as irons and tend to erode if left outside too long. If they are fresh, stonys will often display a blackened fusion crust. Stonys are of two basic classes, the **chondrites** and the **achondrites**. Chondrites have chondrules, little bits of metallic looking silica; achondrites do not.**

The stony-iron type is a mix of the two, and is very rare. Notice there is no "comet" meteorite. The ice of a comet fragment completely burns up, and the small solid pieces in a comet are too small to survive either.

*There are two basic types of meteor ("shooting star") events: sporadic and shower. **Sporadic meteors** can appear anywhere in the sky at any time and travel in any direction. They generally appear alone. On a dark, clear night a typical observer can see five to six sporadic meteors per hour. However, throughout the year there are certain times when many more meteors than normal are visible. This is a **meteor shower**. Shower meteors do not move randomly. A shower meteor can appear anywhere in the sky, but if we trace their motions backward, they appear to diverge, or radiate, from one spot in the sky. We call the point from which they appear to diverge the radiant. We name a meteor shower by the location of the radiant. For instance, one of the best meteor showers of the year is the Perseids, so-called because the radiant is in the constellation of Perseus. The Perseids are visible for about two weeks in August, but they peak around August 13. We can track the motions of incoming meteors to figure out what kind of orbits their parent bodies had around the Sun. Sporadic meteors come from objects that have orbits similar to minor planets—nearly circular orbits that are not too inclined to Earth's orbital plane. Thus, it appears that sporadic meteors come from minor planets. On the other hand, the bodies that produce shower meteors have orbits that resemble comets. In fact, some meteor showers come from bodies that follow the orbits of known comets. What about meteor showers that are not associated with any known comet? They probably come from extinct comets. Remember that comets do not last too long.*

**Today we take meteor impacts on Earth for granted, and some can be dramatic. An 8½-pound meteorite crashed through the roof of a small house in Alabama in 1954, smashed through the second floor and bounced off the living room floor, striking Mrs. Hewlett Hodges and giving her a big bruise on her hip. A 26-pound meteorite smashed in the back end of Michelle Knapp's Chevy in Peekskill, New York in 1992. In 1994, Jose Martin was driving with his wife near Madrid, Spain when a 3-pound rock smashed through the windshield, bent the steering wheel and hit him on the finger, breaking it. However, the odds of you being hit and killed by a meteorite are vanishingly small. Space is big, Earth is a small target, and you are an even smaller one.

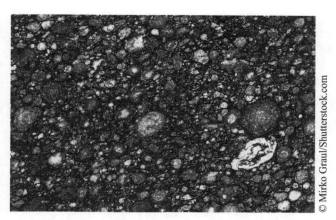

Figure 7.47 Chondrules in a carbonaceous chondrite.

Figure 7.48 Fragment of a meteorite that fell near Chelyabinsk in Russia in 2013.

*"The accepted nomenclature is as follows: the object itself is a **meteor**; the body in [outer space] is sometimes called a **meteoroid**; the fallen body is a **meteorite**.*

*A **fireball** is bright enough to cast a shadow; a **bolide** explodes with an audible noise."*

—Cecilia Payne-Gaposhchkin, Introduction to Astronomy, 1954**

The Sun and the Stars

Chapter 8

> *"The sun is a mass of incandescent gas, a gigantic nuclear furnace where hydrogen is built into helium at a temperature of millions of degrees."*
>
> **—from the song "Why Does the Sun Shine?"**

> *"Can you bind the chains of the Pleiades or loosen the belt of Orion?*
> *Can you bring forth the constellations in their seasons or lead out the Bear and her cubs?"*
>
> **—Job 38:31-32**

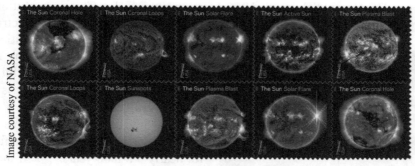

Figure 8.1 2021 Postage stamps highlighting Multi-wavelength images of the Sun from Nasa's Solar Dynamics Observatory

The Sun

At the center of our solar system we find the "greater light" of Genesis 1, enormous, complex and powerful. In most ancient cultures the Sun was revered as a god, but in Genesis 1 we are introduced to the God who transcends His creation. Of all the luminaries, the Sun is the most dominant when it comes to

*From *Learning Astronomy, 3/e* by Wayne A. Barkhouse and Timothy R. Young. Copyright © 2020 by Kendall Hunt Publishing Company. Reprinted by permission.

***From *Fundamentals of College Astronomy, 4/e* by Michael C. LoPresto and Steven R. Murrell. Copyright © 2019 by Michael C. LoPresto. Reprinted by permission.

****From *The Heavens & The Earth* by Marcus Ross, John Whitmore, Steven Gollmer, Danny Faulkner. Copyright © 2015 by Marcus Ross, John Whitmore, Steven Gollmer, Danny Faulkner. Reprinted by permission.

the purpose for these lights in the sky to (1) separate day from night, (2) to help us mark the passage of time, (3) to give light upon the Earth, and (4) to declare God's glory. An important point in the creation perspective is that the Sun is not the primary source of life—God is. However, as our nearby star it lends itself to detailed observations that reveal a great deal about its composition, organization, and activity.

The Sun is "only" 93,000,000 miles away (1 AU), while our next closest stellar neighbor is over 20 *trillion* miles away. The diameter of the Sun is roughly 110 times that of the Earth, and the volume of the Sun is great enough to hold about 1 million Earths. The Sun comprises 99.86 percent of all the mass in our solar system. If we represented the Sun with a 10-pound bowling ball, then everything else in our solar system could be represented by the combined mass of one nickel and one penny. Jupiter would be the nickel.

In many ways, the Sun is just an ordinary star in terms of composition, temperature, and brightness. But in other ways, it is clear that the Sun is designed for life to be possible on Earth. For example, some stars have superflares that release enormous amounts of deadly radiation. The Sun's solar flares are mild. The Sun's temperature and distance from Earth are ideal for life, although many scientists chalk that up to cosmic coincidence. Strangely, the Sun is depleted in lithium by a factor of 100 compared to other similar stars. We have not yet discovered the reason for this, but perhaps it will turn out to be yet another feature of design.

****Astronomers use spectroscopy to determine the Sun's composition. The Sun's spectrum has many dark absorption lines. Different elements produce different lines, so astronomers can study the relative strengths of the lines to measure the composition. The Sun's composition is 73 percent hydrogen, 25 percent helium, and 2 percent everything else. This composition matches that of the Jovian planets well, but not the terrestrial planets. Since we observe the spectrum of only the Sun's photosphere, this composition is that of the photosphere. However, it seems likely that the Sun's gases have mixed quite a bit, so the photospheric composition probably is close to the Sun's overall composition.****

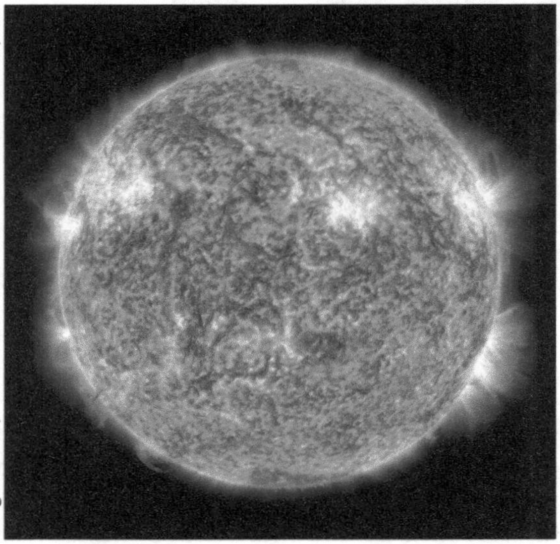

The Sun can create problems for secularists who believe that the Sun has been fusing hydrogen for nearly 5 billion years. ****If this theory of the Sun's energy source is correct, then we would expect that the Sun's brightness ought gradually to increase over time. Computation shows that if the Sun is 4.6 billion years old, the Sun ought to have brightened by about 40 percent. Most evolutionists think that life arose on Earth 3.5 billion years ago. Since that time, the Sun ought to have brightened by 25 percent. A 25 percent increase in solar brightness would increase the average temperature on Earth by 17°C. The average temperature on Earth today is 15°C, so the average temperature of Earth when life supposedly first arose would have been –2°C. If this were true, Earth would have been almost completely

Figure 8.2 The Sun is not absolutely constant in terms of its energy output. However, its variation is of a level that has minimal impact to life on the Earth.

frozen back then, but no one believes this. Evolutionists call this the *young faint sun paradox*. Scientists have put forth a number of explanations for this, but none is agreed upon. However, if Earth is only thousands of years old, this is not a problem.****

The Sun also resists naturalistic formation scenarios. The most popular evolutionary model claims that the Sun (as with other stars) formed by the collapse of a nebula—a giant cloud of hydrogen and helium gas. The model requires many millions of years for this to occur, so of course no one has actually witnessed the process. The outward force of gas pressure in a typical nebula far exceeds the meager inward pull of gravity. As far as we can observe, nebulae only expand and never contract to form stars. Even if gravity could somehow

Cepheid Variable Star V1 in M31 · Hubble Space Telescope · WFC3/UVIS

Image courtesy of NASA, ESA, and the Hubble Heritage Team (STScI/AURA)

Figure 8.3 Unlike the Sun, many stars vary significantly in their energy output due to a variety of reasons. This image taken by the Hubble Space Telescope demonstrates the change in brightness of a Cepheid variable star over the course of a few weeks. If our Sun were a Cepheid, life as we know it on the Earth would be dramatically altered.

overcome gas pressure, magnetic fields and angular momentum would tend to resist any further collapse, preventing the Sun from forming at all. Naturalists do propose a variety of hypothetical scenarios to support their model, but the science is not there to contradict what the Bible teaches. The current contradictions are rooted in naturalism—a philosophy—and not in science.

8.1 General Properties

*The Sun is the most-studied star in our universe for obvious reasons. Information obtained by careful measurements of the properties of the Sun helps astronomers to piece together our understanding of other stars.

One of the most important measurements of the Sun is the amount of energy that reaches the Earth, the solar constant or flux. From Table 8.1, we see that the observed solar constant, extrapolated over all wavelengths, is approximately 1,367 W/m² (Watts per square meter). We can use this information to estimate the luminosity of the Sun given the Sun–Earth distance. The luminosity of the Sun can be calculated by:

$$L_\odot = f_\odot\left(4\pi d^2\right) = \left(1{,}367\frac{\text{W}}{\text{m}^2}\right)(4\pi)\left(1.496\times10^{11}\text{m}\right)^2 = 3.844\times10^{26}\text{W},$$

where $\odot$ the symbol represents the Sun, d is the average Sun–Earth distance (1 AU), and f is the flux.* You can think of the Sun as a 3.844×10^{26} Watt light bulb. Luminosity is an object's intrinsic brightness.

Table 8.1 Basic Physical Properties of the Sun	
Mass	2.0×10^{30} kg
Radius (equatorial)	7.0×10^{8} m
Angular size	31.6–32.7 arc-minutes (about ½ degree)
Density (average)	1.4×10^{3} kg/m³
Rotation period (equatorial)	25.38 days (sidereal)
Surface temperature	5,800 K
Core temperature	15.7 million K (model dependent)
Solar constant	1.367×10^{3} W/m²
Luminosity	3.839×10^{26} W
Absolute magnitude	+4.83
Apparent magnitude	−26.74
Spectral type	G2

The radius of the Sun, $R_{\odot}$, can be determined from the angular size of the solar disk as viewed from the Earth (q ~0.53°), along with the Sun–Earth distance. An estimate of the solar radius is calculated from

$$R_{\odot} \approx 0.5d \tan \theta = 0.5 \,(1.496 \times 10^{11} \text{ m}) \tan (0.53°) = 6.92 \times 10^{8} \text{ m}$$

Once the luminosity and radius of the Sun are determined, we can use an equation from the Stefan–Boltzmann law to estimate the effective surface temperature of the Sun by assuming that it acts like a blackbody (a theoretical "perfect radiator" of energy). The law states that the luminosity L is proportional to the star's surface area and the fourth power of its surface temperature:

$$L = 4\pi r^2 \sigma T^4$$

Reorganizing the equation, this temperature is given by:

$$T_{eff} = \left[\frac{L_{\odot}}{4\pi\sigma d^2}\right]^{1/4} = \left[\frac{3.844 \times 10^{26} \text{W}}{(4\pi)(5.67 \times 10^{-8} \text{W/m}^2\text{K}^4)(6.92 \times 10^{8}\text{m})^2}\right]^{1/4} = 5.79 \times 10^{3}\text{K}.$$

The mass of the Sun can be estimated using Kepler's third law of planetary motion:

$$P^2 = \frac{4\pi^2}{G(M_\odot + M_\oplus)} a^3 \rightarrow M_\odot = \left(\frac{4\pi^2 a^3}{GP^2}\right) - M_\oplus$$

$$= \left[\frac{4\pi^2 \left(1.496 \times 10^{11} \text{m}\right)^3}{\left(6.672 \times 10^{-11} \text{m}^3/\text{kgs}^2\right)\left(3.1558 \times 10^7 \text{s}\right)^2}\right] - \left(5.97219 \times 10^{24} \text{kg}\right) = 1.989 \times 10^{30} \text{kg},$$

where $M_\oplus$ is the mass of the Earth. Don't be frightened by the math. The point of these examples is to show that by using a combination of direct observations and measurements, along with several laws and mathematical equations, we can determine a variety of important properties for an object that is 93,000,000 miles away.

8.2 Structure of the Sun

The Sun, like most stars, consists of hot plasma (ionized gas) that is held together due to the combined gravitational force of its mass (directed inward) and gas pressure (directed outward). The Sun is in a state of constant balance or equilibrium between these two competing forces, in much the same way that a filled balloon maintains a certain size (assuming no air leaks), with the air pressure on the inside of the balloon balancing the atmospheric pressure surrounding the balloon on the outside. We call the results of this balancing act between the inward force of gravity and the outward push of expanding gases the *hydrostatic equilibrium*. In theory, if a star began to cool down or started to heat up, it would reach a new state of hydrostatic equilibrium. Thankfully the Sun has been very stable.

Scientists are able to make great leaps in our understanding of the structure of the Sun and other stars by trying to model their interiors. In Table 8.1, we see that the temperature of the Sun changes dramatically from the surface (~5,800 K) to the center (~16 million K). Understanding how the temperature and other physical properties of the Sun (e.g., density and pressure) change from the surface to the center, allows us to determine whether we have a good comprehension of the physics involved in how stars exist and live out their lives. To model the physical properties of stars, including our Sun, we use a variety of basic equations that relate how physical properties change with radius and see whether they match observations.

With an understanding of hydrostatic equilibrium and various other physical principles, we can begin to understand the inner workings of the Sun in terms of its structure via computer modeling and matching simulation results with observations. *These models are usually tweaked by making minor adjustments in assumptions and physical conditions that go into the models to get as close a match to the real Sun as possible. This enterprise allows us to check to see whether we have a good fundamental understanding of the physics behind how a star works.

One important technique that has been developed over the last two decades has been the use of helioseismology to probe the internal structure of the Sun. The Sun undulates or "rings" like a bell as pressure waves pass through its surface from the interior. These acoustic pressure waves can be used like ground waves on the Earth that are generated by earthquakes, to probe the interior structure of the Sun. Helioseismology has greatly advanced our study of the Sun and provides precision testing of our solar models.*

8.3 Solar Atmosphere

The atmosphere of the Sun generally refers to the layers that lie above the convection zone—the outermost region of the solar interior. The three main components of the solar atmosphere include the photosphere, the chromosphere, and the corona. Each of these regions has distinct characteristics that we will look at in greater detail.

Photosphere

*When you look at the Sun with a projection screen or your protected eyes (beware that staring at the Sun can and will cause permanent eye damage), the disk that you see is that part of the solar atmosphere that we call the photosphere. The photosphere is approximately 500-km thick and marks the region of the atmosphere where visible light photons predominately escape. For depths further down in the Sun, the gas continues to increase in density, thus making it harder for visible light photons to travel to the surface without being scattered and absorbed. Astronomers use the term optical depth to measure the amount of absorption of light. The optical depth has no units associated with it. For the photosphere, the optical depth is on average $t = 2/3$, indicating that visible light photons have a good chance of leaving the surface of the Sun without being absorbed. Since the thickness of the photosphere is only 500 km (0.07 percent of the solar radius), this gives rise to the sharpness of the visible disk of the Sun.

The temperature of the photosphere generally ranges from 6,500 K at the top of the convective zone to approximately 4,400 K at the top of the photosphere. This is consistent with the measurement of the effective temperature of the Sun of 5,800 K since it is from this layer of the solar atmosphere that the visible photons originate that strike the Earth.*

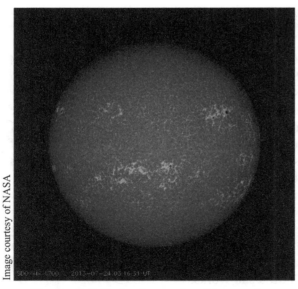

Image courtesy of NASA

Figure 8.4 Solar Dynamics Observatory ultraviolet image of the Sun's chromosphere.

Chromosphere

*The layer of the atmosphere that lies directly above the photosphere is the chromosphere. The chromosphere is approximately 2,000- to 3,000-km thick and is transparent to visible light. This makes the chromosphere difficult to observe directly when viewing the Sun. In general, the chromosphere is observed during a solar eclipse when the Moon blocks the light from the photosphere. This allows the chromosphere to be viewable near the edge of the solar disk against the black background of space. The color of the chromosphere is a reddish hue due to the recombination of electrons in hydrogen atoms that have been ionized.

Viewing the Sun in non-optical wavelengths also allows us to directly image the chromosphere. Figure 8.4 shows the Sun's chromosphere as viewed at ultraviolet wavelengths using the NASA space-based *Solar Dynamics Observatory*.

The temperature of the chromosphere, in contrast to the photosphere, increases from approximately 4,400 K at the photosphere/chromosphere boundary to roughly 10,000 K to 25,000 K at the top. The density also drops with increasing height throughout this part of the solar atmosphere, with typical density values 10,000 times less than that in the photosphere.*

Corona

The third main component of the Sun's atmosphere is the corona. The corona sits above the chromosphere and a smaller region called the transition layer, and extends for several million kilometers into space. The corona is best observed at visible wavelengths during a solar eclipse when the Moon blocks the intense light from the photosphere (see Figure 8.5). Spectroscopic analysis of the light from the corona during a solar eclipse shows the signature of highly ionized elements, including calcium and iron. The temperature of the corona increases to several million Kelvin as evident by the emission signatures from ionized metals. Such high temperatures are consistent with the observed x-ray emission from the Sun. The density of the corona is lower than that of the chromosphere and is on average about 10^{15} particles/m^3. Due to the low density, the amount of heat generated in the corona is low even though temperatures are high.

The cause of the high temperature of the corona has been a mystery for many decades. Naively you would think that the further away from the central heat source in the core of the Sun the lower the temperature. However, the temperature of the Sun's atmosphere increases to millions of degrees before the temperature in the outer corona begins to decline. How is this possible? One possibility is that magnetic fields are somehow able to release energy to heat the gas in the solar atmosphere. Details of this process are not yet fully understood.

Image © psamtik, 2014. Used under license from Shutterstock, Inc.

Figure 8.5 The Sun's corona is best viewed during a solar eclipse when the Moon passes directly in front of the Sun as viewed from the Earth.

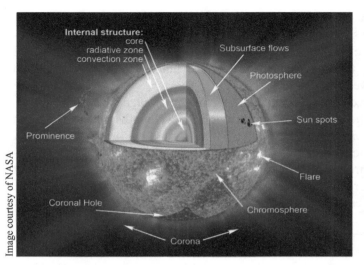

Image courtesy of NASA

Figure 8.6 This cutaway diagram illustrates the main components of the Sun's internal structure and its atmosphere.

8.4 Solar Interior

The interior of the Sun contains the furnace that is responsible for the generation of the heat and light that sustains life on the Earth. Both the radiative and the convection zones are found inside the Sun, all of which transport energy released in the core in the form of gamma-rays and x-rays to the surface.

The Core

Before the discovery of nuclear fusion, the mechanism responsible for the production of energy in the Sun was unknown. Ancient cultures assumed that the Sun was made up of some type of special material that would burn forever. Starting in the mid-1800s, several eminent scientists made suggestions of the possible source of solar energy using modern physical laws. John Waterston showed that the Sun would only last for approximately 20,000 years if it was burning fuel by a standard chemical or combustion process (imagine the Sun composed of a sphere of gasoline or coal). In 1854, Hermann Helmholtz proposed the gravitational contraction of the Sun as a way to generate energy. Lord Kevin in 1887 modified the gravitational collapse scheme. However, it would not be until the twentieth century, with breakthroughs in our understanding of the atom and nuclear reactions that the puzzle would be solved.

Nuclear Fusion

*In 1938, Hans Bethe (awarded the Noble Prize in physics in 1967) published a pair of papers that described the energy source that powers the Sun. Like other normal stars, the Sun is the hottest at the center or core region. Einstein's theory of special relativity in 1905 and the publication of his famous equation $E = mc^2$ can be used to understand how the Sun generates its energy. The thermonuclear fusion of hydrogen into helium in the core of the Sun is responsible for the production of solar energy. During this process, mass is converted into energy as hydrogen is converted into helium. Einstein's $E = mc^2$ equation tells us how much energy is released for a given amount of mass. Since the mass term is multiplied by the square of the speed of light, a small amount of mass will produce a huge amount of energy.

There are several steps involved in the hydrogen fusion process (proton–proton cycle) occurring at the center of the Sun.* The simplest form of hydrogen (H-1) is made up of just one proton, so the name proton–proton cycle (or proton–proton chain) is actually referring to hydrogen nuclei. Deuterium is an isotope of hydrogen that also has just one proton, but also one neutron for a total mass of 2. Helium nuclei must have two protons, but may be in the form of He-3 (2 protons and 1 neutron) or He-4 (2 protons and 2 neutrons).

$$\,^1_1\text{H} + \,^1_1\text{H} \rightarrow \,^2_1\text{H} + e^+ + \nu_e$$

(Formation of deuterium; hydrogen-2 ν is a neutrino)

$$\,^2_1\text{H} + \,^1_1\text{H} \rightarrow \,^3_2\text{He} + \gamma$$

(formation of helium-3 from one deuterium and another proton)

$$\,^3_2\text{He} + \,^3_2\text{He} \rightarrow \,^4_2\text{He} + \,^1_1\text{H} + \,^1_1\text{H}$$

(formation of helium-4; the two H nuclei then keep the chain reaction going)

In the end, there is a tiny difference in mass as protons are transformed into neutrons. This "missing" mass was converted into energy, as per Einstein's famous equation, $E = mc^2$. It takes very little mass to produce a great amount of energy.

Until recently, astronomers were puzzled by what was known as "the solar neutrino problem." *The neutrinos produced during the proton–proton cycle were believed to be very low mass and could be taken as massless (like a photon) and thus stream out from the solar core at the speed of light, taking only 8.3 minutes to reach the Earth. Calculation of the expected number of neutrinos detected on Earth from various experiments showed a mismatch between what was actually detected and what was predicted. Essentially one-third of the number of expected neutrinos were detected from the Sun. Many suggestions were made to explain this mismatch between theory and observations. In the end, it was discovered that neutrinos are not massless particles and that they "change flavors" (one of many bizarre terms used by physicists) in route from the Sun to the Earth. Early detectors used to observe neutrinos from the Sun were only sensitive to one type of neutrino flavor. Since there are three types of neutrinos and the neutrino can change flavor (change form), an average of only one-third of the total number of neutrinos

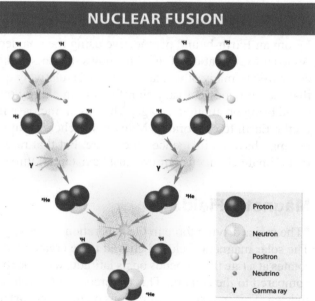

Figure 8.7 The proton-proton fusion cycle is illustrated here.

Image © Designua, 2014. Used under license from Shutterstock, Inc.

would be detected. Thus, the flavor-changing property of the neutrino, known as "neutrino oscillation," was discovered by making observations of the Sun.

*In the first reaction of the proton–proton chain, two hydrogen atoms are converted into deuterium, a heavier form of hydrogen. One of the two protons is converted into a neutron, releasing a positron and a neutrino in the process. For this process to occur, two protons must be brought very close together. Since each proton is positively charged, there is a strong electric force (Coulomb force) pushing these objects apart since the charges are of the same sign (recall that charges with the same sign repel while opposite charges attract). The extreme high pressure and temperature found at the center of the Sun allows atoms to overcome their electrostatic repulsion. The incredible density of the core assures that there will be enough particle collisions in the process. The minimum temperature required for proton–proton nuclear fusion is on the order of 10 million K. The Sun has an estimated core temperature of 16 million K (see Table 8.1).

The short-wavelength gamma-ray and x-ray photons that are generated in the center of the Sun do not travel directly to the solar surface. Due to the high density in the core, the average photon released in the proton–proton cycle travels approximately 0.1 cm (mean free path) before being absorbed or scattered.* As photons work their way toward the surface, the collisions convert some of the high-energy photons into lower-energy radiation. The region where this occurs is referred to as the radiative zone. As energy from the core eventually reaches the outer layers of the Sun's interior the hot gases begin to churn in convection currents, with hotter gas rising and cooler gas sinking, so astronomers call this region the convection zone. The convection zone brings energy to the surface, bubbling up at the photosphere.

8.5 Solar Activity

From an Earth-bound perspective using the unaided eye, the Sun seems eternal and unchanging. A closer examination of the Sun shows that this is far from the truth. The Sun is an active star, but compared to many other stars, the Sun is relatively quiet. The presence of life on the Earth requires that the Sun does not vary significantly in energy output. A decrease in the luminosity of the Sun could bring on another ice age, whereas an increase in the energy output could cause all the oceans on the Earth to evaporate. Many researchers acknowledge that even minor variation in solar output may have an influence on climate, but that rarely becomes a talking point for those who feel that climate change is our greatest "existential threat."

Magnetic Field

*The major driving force in the generation of activity on the Sun is the presence of a magnetic field. The solar magnetic field is believed to be created from the differential rotation of the Sun. The Sun rotates at a rate that depends on its latitude, with the rotation period being longer near the rotation axis compared to the equator. This difference in the rotation period causes the magnetic field to become twisted. This in turn creates regions on the solar surface where loops of magnetic field burst through, releasing huge amounts of energy. The "Babcock Model" attempts to explain the magnetic cycle of the Sun as being the result of this repeated tangling and untangling of the Sun's magnetic field.

The solar magnetic field undergoes an eleven-year solar cycle in which the magnetic field reverses polarity every eleven years. The magnetic activity of the Sun changes such that at certain times the Sun goes through a period of solar maximum (greater than average magnetic activity) and solar minimum (period of low magnetic activity). The time between consecutive solar maximums is approximately eleven years. During each solar maximum, the Sun changes the polarity of its magnetic field such that the north magnetic pole becomes a south magnetic pole and vice versa. This magnetic polarity then switches back after another eleven years have elapsed during the next solar maximum.*

Sunspots

*One of the important observations that Galileo made with his telescope was that the Sun has dark spots or patches on its surface. These sunspots (Figure 8.9) reside on the photosphere and are associated with magnetic activity that erupts on the solar surface from the interior. Sunspots appear dark because they have a lower temperature than the surrounding solar atmosphere. Typically, sunspots have temperatures ranging from 3,000 to 4,500 K, while the temperature of the photosphere is approximately 5,800 K. The darkest (and coolest) region of a sunspot is called the umbra, the same term used for the darkest part of a shadow in an eclipse.

Sunspots are transient phenomena in the sense that they form, enlarge in size, and then dissipate after several weeks or months on average. Sunspots both move across the surface of the Sun with their own proper motion and are also carried along by the Sun's differential rotation. Some spots have long lifetimes so that they travel behind the Sun as viewed from the Earth and reappear on the opposite side after a couple of weeks. Sunspots can range in size from as small as

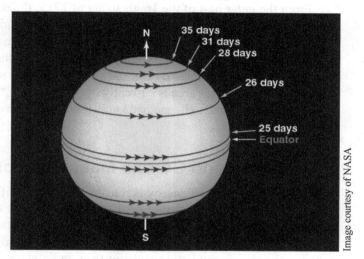

Image courtesy of NASA

Figure 8.8 Since the Sun is not solid, it experiences differential rotation, rotating faster at the equator and slower at the poles.

Image courtesy of NASA/SDO/AIA/HMI/Goddard Space Flight Center

Figure 8.9 Solar Dynamics Observatory image of the Sun showing sunspots (dark patches) and other solar activity near the edge of the solar disk.

only a few kilometers in diameter to many times larger than the Earth.* Sunspots usually appear in pairs and show a magnetic field that is much stronger than the Sun's average magnetic field.

*Since sunspots are directly connected with the magnetic field of the Sun, sunspots also go through an eleven-year solar cycle like the magnetic field. When the Sun is at solar maximum, the number of sunspots is greater on average than when the Sun is at solar minimum.

From the latter half of the 1600s to about 1715, the number of sunspots visible on the Sun was greatly reduced. This time period is known as the Maunder Minimum, and people have suggested that the reduced sunspot activity corresponded to abnormally cold weather on the Earth at that time (part of "the Little Ice Age" of roughly 1300 to 1865). Thus, correlations between solar activity and the climate on the Earth have been the subject of many studies, with no real concrete connections having been made. This area is currently open for debate.* Despite some of the more extreme views today regarding climate change, we know there have been periods of warming and cooling throughout recorded history.

Solar Flares and Corona Mass Ejections

*Occasionally, a burst of energy is released from a specific area of the Sun. The event, known as a solar flare, can release up to one-sixth of the total energy that the Sun generates in one second from nuclear fusion. This tremendous amount of energy heats up the overlying atmosphere and causes ionized gas to be expelled into space. The radiation emitted by solar flares covers the complete range of the electromagnetic spectrum from radio waves to gamma-rays.

The direction that a solar flare takes when it erupts from the surface of the Sun can impact the environment on Earth. Energetic particles that get accelerated by the flare can interact with the Earth's atmosphere and cause a radiation hazard to astronauts or people in high flying airplanes. Satellites in space can also be in danger from short circuits and overloaded electrical equipment as the density of charged particles greatly increases in the vicinity of the spacecraft.

Solar flares are usually generated near sunspot groups and are thus related to the activity of the solar magnetic field. The exact mechanism that triggers a solar flare is not known. Some of these flare produce very powerful bursts of x-rays, and they are one of the most energetic events on the Sun.

Often associated with solar flares are coronal mass ejections (CMEs). CMEs are bursts of charged particles, usually electrons and protons, from sites of strong magnetic field activity. The particles

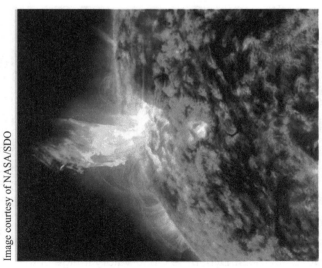

Image courtesy of NASA/SDO

Figure 8.10 The Solar Dynamics Observatory captured this solar flare on June 20, 2013.

emitted during a CME have speeds ranging from 200 km to 1,000 km per second. Usually these regions are correlated with the presence of sunspots on the photosphere. There is also a greater than average probability that the eruption of a solar flare will be followed within a few hours by a CME, but the link is not clearly established. A solar flare is more powerful than a CME in terms of the energy that is released during the event.

CMEs send out particles from the Sun into interplanetary space. Occasionally, the alignment is such that the emitted charged particles will intersect the Earth and interact with the magnetosphere. This in turn can cause communication disruptions, power grid failures, and beautiful displays of the northern and southern lights.*

Granulations and Prominences

As described earlier, the energy generated in the core of the Sun via nuclear fusion of hydrogen is carried to the solar surface through radiation and convection. Since the convective zone is just below the photosphere, the convection cells that transport the energy upward rise up to the photosphere and then fall downward as they release their heat and lose their buoyancy. The tops of the convective cells are visible on the photosphere as a grainy or cellular-like pattern referred to as granulation (Figure 8.12).

*Individual cells are on average about 1,000 km (over 600 miles) in diameter and last anywhere from ten to twenty minutes in duration. Motion of granulation cells can be measured via the Doppler Effect as cells rise up from the solar interior. The observed motion of the cells and their duration matches that expected for the convection process.

Image courtesy of NASA/STEREO

Figure 8.11 The NASA STEREO spacecraft captures an image of the Sun during a coronal mass ejection (right side). This observation was recorded on May 20, 2011.

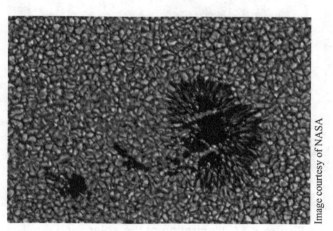

Image courtesy of NASA

Figure 8.12 Solar granulation pattern represents the movement of convection cells from the interior to the surface of the Sun.

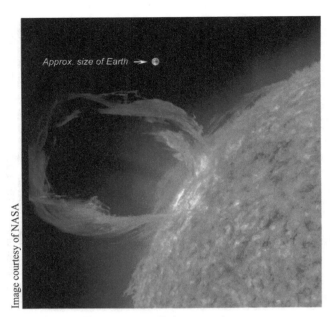

Image courtesy of NASA

Figure 8.13 Solar prominence.

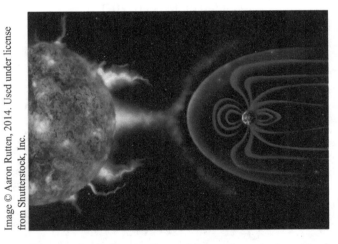

Image © Aaron Rutten, 2014. Used under license from Shutterstock, Inc.

Figure 8.14 Illustration of the solar wind emitted by the Sun and its interaction with the Earth's magnetosphere. Note that the illustration is not to scale.

The photosphere is also home to loops of ionized gas or plasma that follow magnetic field lines that extend out into the corona. This phenomenon is known as a prominence and is characterized by loops of plasma that extend outward for hundreds of thousands of kilometers from the photosphere. Prominences can form over timescales of days and can last for several weeks to months. In addition to being associated with magnetic fields, prominences are also related to CMEs.*

Solar Wind

*The solar wind is a stream of charged particles that are released from the outer atmosphere of the Sun. The density, temperature, and speed of the solar wind change with time, and these properties are related to the solar cycle. During a corona mass ejection, large bursts of solar wind particles are emitted from the Sun. The solar wind spreads throughout the solar system in a bubble-like pattern that engulfs all the planets and extends far beyond the orbit of Pluto. The amount of mass that the Sun loses per year due to the solar wind is approximately 10^{-14} solar masses. This is a negligible amount, although some stars lose a significant amount of their mass via strong stellar winds.

The impact of the solar wind on the Earth can cause problems when it leads to geomagnetic storms. Energetic charges released during solar flares and CMEs can get trapped in the Van Allen radiation belts that surround the Earth. Normally these solar wind particles are deflected around the Earth by its magnetic field, but especially energetic particles released during high solar activity can punch their way into the Earth's magnetic field.

One of the most dramatic results of this type of interaction of the solar wind particles with the Earth's atmosphere is the production of the aurora borealis and aurora australis (northern and southern lights). The charged solar wind particles travel along the Earth's magnetic field lines from the north to the south magnetic poles. As they move between poles, they collide with atoms in the Earth atmosphere, which in turn ionizes the gas. As electrons recombine with the ionized atoms, photons are released. These photons are visible from the Earth as the northern and southern lights.*

Figure 8.15 The northern lights or aurora borealis are caused by solar wind particles ionizing atoms in the Earth's upper atmosphere. As electrons recombine with the ions, they release photons that we see as the northern lights.

8.6 Understanding other Stars: Brightness

We know now that the Sun is just one of billions of stars in our Milky Way galaxy, and the Milky Way is just one of perhaps billions of galaxies. That is a lot of stars! *Standing on Earth looking up at the nighttime sky, we see a few thousand stars. With a pair of binoculars pointed upward, a smaller patch of sky becomes littered with stars. Then, with a telescope in a smaller piece of sky, the field of view is lit up with stars.

How bright are stars? Hipparchus in 172 BC set up a comparison system to measure the brightness of stars. Recording and magnifying devices had not been invented by this time, so his only instrument was his eye. Hipparchus took the brightest star in the sky and called it a first class star or first magnitude; the next brightest was second class or second magnitude and so on, ending at sixth class. This made sense because he believed that the brighter ones were closer, meaning they all intrinsically

Figure 8.16 A stamp printed in Greece shows ancient Greek astronomer Hipparchos and his astrolabe.

had the same brightness. He also found variation within the classes and wrote about how none were the same brightness, or "bigness or lustre," as he wrote it. The fainter a star, the larger its class or magnitude becomes. In our modern study of stars, this seems to be reversed. It might seem more intuitive to have fainter stars assigned smaller numbers, but the general system devised by Hipparchus has been accepted as the apparent scale for magnitudes. The limit of the human eye in ideal dark conditions is a class 6 or sixth magnitude and can be written in more compact form $m = 6$, as can be for any star.* Due to light pollution, it would be rare for someone to see a naked-eye object today at $m = 6$. As light pollution increases, observers no longer see $m = 5$ objects, and then $m = 4$ objects, etc. Depending on where you live, you could "lose" most of your stars.

The human eye is an incredibly adaptive instrument, so much so that it can see in very low light and in full daylight. In order to do this, the eye must start to drop light particles called photons, as the object gets brighter. Because the eye discards more and more photons in increasingly brighter light the eye counts photons logarithmically. Using the eye as an instrument is fine if we are doing rough calculations, and many amateur astronomers do just eye observations. But we know a better way is to measure the brightness of light, called the intensity of the light. The intensity is measured not logarithmically but linearly, so it makes it easier to compare star brightness. The relationship between intensity and magnitude is such that a difference in magnitude of 5 is an intensity difference of 100. (This was discussed in Chapter 2—for convenience we said each order of magnitude resulted in an intensity difference of roughly 2.5).

*A CCD image of a star field shows bright and faint stars. The stars are all just points of light because they are far away, but because the light from the bright stars "spills" into areas on the CCD chip the star appears bigger. In fact some of the bright ones may be smaller in actual stellar radius than the faint ones. You need to get more information before the size of the star can be determined.

For example, the star Antares has magnitude $m = 1$. By comparison a star with magnitude $m = 6$ would appear fainter than Antares by a factor of 100 times. The exact formula that can be used to compare any two magnitudes and convert to a brightness ratio is:

$$\frac{I_a}{I_b} = \left(2.512\right)^{(m_b - m_a)}.$$

I^a is the light intensity (brightness) and m^a is the magnitude of star a, and likewise I^b is the light intensity (brightness) and m^b is the magnitude of star b.

See chapter two to review how we were able to convert from differences in brightness (flux) to differences in magnitude, and vice versa. This equation is just another way of showing these same relationships we discussed and calculated in chapter two.

The magnitude scale can go to negative numbers, so it can be confusing. For example, a star with $m = -2$ is brighter than stars with $m = -1$ or $m = 4$.*

Magnitude:	−26.75	−12.9	−4.89	−2.94	+1.04	+1.16
	Sun	Moon	Venus	Jupiter	Spica	Formulahault

*At a time when many phenomena were being explained in the late eighteenth century, light, as well as other concepts such as electricity, magnetism, and gravity, all shared the physical aspect of decreasing intensity with increased distance. The further you are away from something, the less the intensity or force of that object. So it is an inverse relation, and not only that, but the distance is taken to the power 2. This is something that many are familiar with as the inverse square law for light: *

$$I \propto \frac{1}{d^2}.$$

The intensity is inversely proportional to the distance squared. If the distance is doubled, the result is one-fourth the intensity. If the distance is tripled, the result is one-ninth the intensity, and so on. If the distance is cut in half, the intensity goes up by a factor of 4.

*The reason light energy is less at a greater distance is that it is spread out over a larger surface area. It is diluted or dimmed at larger distances. A source of light distributes the energy in all directions and streams light particles directly away from the source. If we had a shell around the source, we could capture the light and add up its total energy. If we made the shell twice as big and did the same thing, the total energy would be the same. But, at each location on the shell, the light would be dimmer, and the original light source would look faded. At each location on the bigger sphere we are intercepting less light. Adding up all the light from those locations would result in the total energy being the same. Thus, from one shell to any shell at any distance away, the total energy is the same, conserving the energy balance.

Image courtesy of NASA

Figure 8.17 Telescope image of stars in the sky, showing the many different brightnesses.

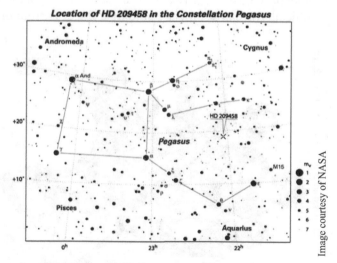

Image courtesy of NASA

Figure 8.18 Star Chart of Pegasus showing star brightness represented by dot size.

Hipparchus required the stars to be at different distances. At the time, it was hard to prove, and even 1,700 years later, the technology had not been accurate enough. Even Tycho Brahe and his suite of instruments, hadn't found the distances to stars. But, the search for distance led to a precision of measuring angles. Measuring angles started centuries ago with the Harappan and Babylonian cultures. They even used triangles in about 3000 BC for looking at stars. Being precise was the difficult part that came later with Greek culture and Archimedes. The definition of 1 degree of measurement in angle is obtained by dividing a circle into 360 angles. Thus 1 degree is 1/360 of a revolution around a circle. This can be further divided into a minute of an arc, called an arc-minute, by dividing a degree by 60. This continues to a second of an arc, called an arc-second and defined as an arc-minute divided by 60. Amazingly, there are $60 \times 60 \times 360$ or about 1.3million (exactly 1.296×10^6) arc-seconds in a circle. That seems like a lot of divisions but there is a lot of space in the sky. The easiest way to measure something in the sky is to take two sticks and put two ends together. This is called the apex. One stick can be pointed at a star, and the other stick can be pointed toward another star or landmark. The amount measured is the distance the sticks are opened by measuring the angle between them. The apparatus is called a sextant (see Figure 8.19) and used widely to measure angles. In combining angles and geometry, one can get distances. It is extremely useful in surveying and architecture since it is impossible to have rulers for large distances.*

*This was the first step in understanding the distance of stars. The method of looking for distances to the stars was very discouraging for Tycho Brahe. He took the best observations possible at the time and found no parallax among the stars. Parallax is something we use every day but very subconsciously. Our two eyes are separated by a distance and used to form an angle of sight on an object. You can see this by holding a finger in front of you and blinking one eye at a time and seeing that your finger moves with respect to a distance object. The further your finger is away from your eyes, the less it moves. Tycho Brahe watched for this movement in the stars. In order to get the separation (like the distance between your eyes) he waited for Earth to move in its orbit for six months. This gave the largest distance possible for parallax observations, even for today's astronomers. He did not see any change in the star pattern and thus concluded that the stars were not at different distances away.

It wasn't until modern times and the invention of the telescope that parallax among stars was seen. The nearest stars show the largest parallax, and the value continues to decrease the further away a star is. For example, Alpha

Figure 8.19 A sextant, used to determine the angle between stars, indicating their positions.

Centauri is 4.3 light years away and has a parallax of roughly 0.75 arc-seconds. Now we see why Tycho had a hard time seeing that small of a number. With the use of a satellite named Hipparcos that was launched in 1989 and took data until 1993, the parallax of 2.5 million stars has been measured.

Interestingly, there is another distance ruler that can be derived from this concept. If we were in a spaceship and backed out perpendicular to the plane of the solar system and measured (with a sextant!) the angle of the Earth at two opposite seasons, we would see that those two positions would appear to get closer and closer as we backed out further and further. If we backed out to the distance in which the angle was one arc-second, we would have a special distance as identified by astronomers. It is called a parsec, and one parsec is equal to exactly $3.08567758 \times 10^{16}$ m.* That is 3.26 light years, which is about 20 trillion miles.

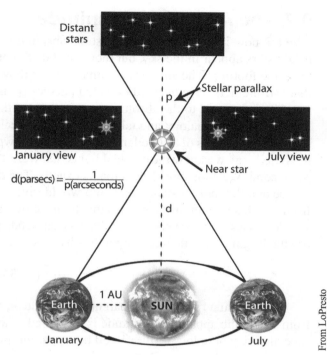

Figure 8.20 Using stellar parallax to measure the distance to a star.

A simple formula that shows the relationship between distance and parallax is

$$d = 1/p$$

where "d" is the distance expressed in parsecs, and "p" is the parallax angle in arc-seconds. As "p" gets smaller, the "d" gets bigger. If we go back to our example of Alpha Centauri, we calculate the distance by taking "$d = 1/0.75$," which gives us $d = 1.3$ pc, which converts back to 4.3 light years.

****Trigonometric parallax is the only direct way to measure stellar distances, but there are many other indirect ways. One such method is the use of **Cepheid variables**. Many stars are variable stars, stars that vary in brightness. Cepheid variables are one type of variable stars. Cepheid variables change brightness in a very regular way over a period that can be anywhere from two days to two months. A century ago, astronomers discovered that brighter Cepheid variables had longer periods. This established a period–luminosity relation, a relationship between period and average brightness. Knowing the period–luminosity relation, an astronomer could measure the period of a Cepheid variable to determine the Cepheid variable's absolute magnitude. Knowing the average apparent magnitude, one can compute the distance. There are many other distance determination methods. Most of them rely upon finding a star's absolute magnitude, and then a simple comparison to the apparent magnitude yields the distance.****

8.7 The Absolute Magnitude

*We are now in a position to understand absolute magnitude. We have already talked about how bright stars appear in the sky, but they are at different distances. So, how can we compare them on the same footing? The answer is to imagine that they are at the same distance. A standard distance that astronomers have decided on is ten parsecs (pc). It is a special distance at which the apparent magnitude is equal to the absolute magnitude, and thus the definition of the absolute magnitude.

We make an analogy and take a conceptual break to make it more understandable. If we measured the brightness of a streetlamp several blocks away, we would be measuring its apparent brightness. Let's say we knew the city had lights that were 1,000 Watts. A Watt is a unit of power or Joules per second. We know how bright that is exactly. The difference in brightness between a 1,000 Watts source and the more distant streetlamp would give us a measure of how far the streetlamp is away from us. This is again from the idea that light follows an inverse square law. The brightness or intensity decreases as the inverse of the distance squared, $1/d^2$. This is the same for star light and we can use the brightness scale that Hipparchus had invented to get the distance–magnitude relation:

$$m - M = 5 \log_{10} d - 5.$$

This formula relates the apparent magnitude m, the absolute magnitude M, and the distance d from where the apparent magnitude is measured to the star. When $d = 10$, we can see that this is a special case where $\log_{10}(10) = 1$ and the apparent magnitude is equal to the absolute magnitude.

$$m - M = 5\log_{10}d - 5$$
$$m - M = 5(1) - 5$$
$$m - M = 5 - 5 = 0$$

Or, $m = M$.

This formula can be used when you are given m and M and want to find the distance to the object. Or, if any two of the variables are known, you can find the third.*

Example: Polaris is 133 pc from Earth and has an apparent magnitude of 2.0, so what is its absolute magnitude?

$$2.0 - M = 5(\log(133)) - 5$$
$$2.0 - M = 5(2.12) - 5$$
$$2.0 - M = 5.6$$
$$M = -3.6$$

So, if Polaris was just 10 pc from Earth, it would look remarkably bright!

For a listing of the twenty brightest stars and the twenty nearest stars, see Appendix C.

Appendix C will include apparent magnitudes, absolute magnitudes, distances, and spectral class for each of the stars listed.

Appendix C

8.8 Colors and Temperature of Stars

When talking about the brightness, it is also known that stars have some hint of color to them. How can stars have color? Wein's law tells us that a hot gas of absolute (Kelvin) temperature T will emit most of its light at a certain wavelength. The peak wavelength (color) is inversely proportional to the absolute temperature (higher temperatures result in shorter wavelengths). *In addition, we remember that the spectrum in the visible is given by ROYGBIV. If more photons are emitted in the red part of the spectrum, the star will look red. It is true that stars with temperatures around 3,800 K will appear redder, like Betelgeuse and Antares. And on the other side of the spectrum, the hotter stars of 10,000 K will emit most of their light in the blue, like Spica and Regulus.* Applying Wein's law, we have a way to determine temperatures of distant objects without needing to visit with a thermometer. Even without digging into the math, we can get a general idea of temperature based on color.

*Colors of stars are determined by the wavelength of the photons that are being produced the most. This can be seen from counting the number of photons according to wavelength. The resulting number versus wavelength graph is called a distribution function. It is a curve similar to a bell curve, with fewer photons at higher and lower wavelengths. The peak can shift depending on the temperature of the gas in the surface of the star. More red photons are produced for cool stars and more blue photons are produced for hotter stars. So hot stars are blue and cool stars are red. What is happening is that the temperature of the blackbody, that is, the collisions of the atoms, is controlling the peak production of light and shifts it according to temperature. The light produced is also governed by the Stefan–Boltzmann law, which relates temperature, surface area, and the total energy being emitted by a star. This means that not only is the peak shifted but the total amount of energy is greater for higher temperatures. This physical property of light is very useful to astronomers, because they can use this property to find out the temperatures of stars.

Atoms Can Tell the Temperature of Stars

*We are now in a position to link together several concepts—the atom, its electron structure, its temperature, and the observed spectrum. We know that the neutral hydrogen atom has one electron orbiting in an energy level. Depending on the temperature of its surroundings, it will be bumping into other atoms, and the bumping might have enough energy to give the electron a boost into a higher-energy level. After a split second, it falls down, and that energy is released as

Source: NASA, ESA, and the Hubble SM4 ERO Team

Figure 8.21 It is easy to pick out various colors of stars inside this globular star cluster.

a photon of light. With enough bumps, meaning the temperature is high enough, the electron can permanently be in the higher-energy level. So, the temperature of the gas influences the number of electrons in any given energy level. Kirchoff's third law states that if a cooler gas is in front of a blackbody, it will absorb photons and send them in another direction, leaving an absence of photons of that color, coming from the original direction.

Let's say we have hydrogen at 10,000 K in the atmosphere of the star. Its electrons will be in the second energy level. Light, at all wavelengths from the blackbody, will stream through the slightly cooler hydrogen gas but only certain wavelengths will interact with the atoms of hydrogen. The electrons in the second energy level will absorb only certain wavelengths of light corresponding to the energy level structure of the hydrogen atom. These are well known to be at 656.3 nm (red), 354.3 nm (blue), and two violets. This is called the Balmer series. In the spectrum, we would see dark lines at those wavelengths, called a dark line or absorption line spectrum. The number of atoms at 10,000 K will determine how dark the line is. If the temperature of the gas is lower than 10,000 K, then there will be fewer atoms with electrons in that level, and the line will not be as dark. The electrons will drop down and spend more time in the lower-energy level and thus start to populate the first energy level. If the hydrogen gas is hotter than 10,000 K, then the number of atoms with electrons in the second energy level will depopulate, and the absorption line will become fainter. The electron will be in an upper energy level or ionized. Ionization occurs when there is enough energy for the electron to overcome the attraction to the positive charge of the nucleus and it can leave the atom.

By just looking at the strength of the hydrogen absorption line, astronomers can determine the temperature. So far we have only talked about hydrogen but there are several elements that can also be used, including helium, calcium II, and iron. They each show different line strengths for different temperatures. So with many lines, it is fairly easy to determine a very precise temperature. This analysis of spectra of stars was first done before astronomers knew that the temperature of the gas was the reason for the absorption lines. A whole system was created to organize stars based on the spectrum, and it is called the *spectral type*. Sometimes called the Harvard classification system, it uses letters and numbers and goes from the hottest stars as O stars and decrease to the coolest stars as M. The whole series is O, B, A, F, G, K, M. You can remember this with the mnemonic, *Oh Be A Fine Gorilla Kiss Me.** Numbers are used within a given spectral type to further distinguish stars by temperature. For example, among "G" stars, a G-0 would be the hottest, and a G-9 would be the coolest. After G-9, the next cooler stars would be K-0.

8.9 The Hertzsprung–Russell Diagram

*Observing the spectrum is a good tool to use to get basic data from stars. So now we have two excellent physical properties for categorizing stars, the luminosity and the temperature. In the early 1900s, Danish astronomer Ejnar Hertzsprung and American astronomer Henry Russell, plotted numerous stars on an *x-y* graph using temperature and luminosity. The plot was made such that the *x*-axis increased in temperature to the left. Higher-temperature stars were near the origin, and cooler-temperature stars were toward the right side. Luminosity increased, as one would expect, going up the *y*-axis. They found something very interesting. Most stars formed a band ranging from the lower right to the upper left. This correlation showed that hotter stars are brighter and cooler stars are fainter. They called this band the *main sequence*. They also found a group of stars above the main sequence but to the right and redder—they called these the red giant stars. A group of stars that are even brighter were found and called supergiant stars. The last group was under the main sequence—they are hotter and fainter, called white dwarfs.*

***Any star that is directly above or below another star on the HR diagram will have the same temperature and, therefore, the same

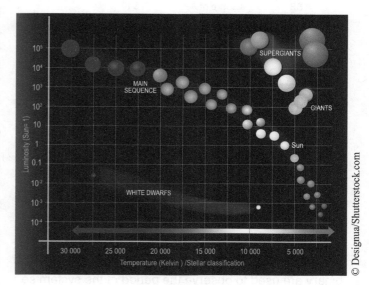

Figure 8.22

color. Any star that is directly right or left of another star has the same luminosity or absolute brightness, but not the same size. Stars of the same luminosity on the right are always larger than those on the left. The reason for this is not hard to understand.

A blue giant is hotter than a red giant, but if both are at the same height on the HR diagram, they have the same luminosity. In order to be as luminous, to give off as much energy, a cooler red giant has to be bigger than the hotter blue giant. This works for dwarf stars too. If a white dwarf and red dwarf are at the same height on the diagram, they must have the same luminosity. A white dwarf can be smaller than a red dwarf but still be as bright because the white dwarf is hotter. Or, stated another way, the cooler red dwarf has to be larger than the hotter white dwarf in order to give off as much energy and, therefore, be of the same luminosity as the white dwarf.***

8.10 Stellar Masses and Diameters

***Being able to determine the mass of stars is very important to astronomers, but how do we determine the mass of a star?

The majority of stars in our galaxy are not alone. Most stars are found in gravitationally bound groups of two or more. When two stars are in orbit of one another, they are called a *binary system*. For example, Sirius is actually a two-star system. To the unaided eye, a binary may look like a single star, but when viewed through a telescope, the two individual stars can be seen. This is called a *visual binary*. It is also possible that the stars are so far away that they will still appear as one when viewed through a telescope, but as they orbit one another, the apparent brightness of the system will

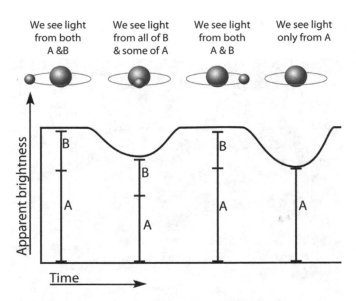

We see light from both A &B · We see light from all of B & some of A · We see light from both A & B · We see light only from A

Figure 8.23 Variations in the brightness of an eclipsing binary are used to observe the period of the system so the masses of the stars can be calculated.

change as the stars alternately block out some of one another's light, as shown in Figure 8.23. This is known as an *eclipsing binary*. In a *spectroscopic binary*, as the name suggests, the presence of the individual stars can only be detected by viewing the lines of their absorption spectra.***

***A binary system should not be confused with an *optical double*. Two stars can appear to be in the same place in the sky from the point of view of Earth because they lie along the same line of sight. The way to tell a double and a binary apart is that the stars in the binary system will be about the same distance away, whereas the two stars seen as a double will be different distances from the observer and could actually be very far from one another.

If the orbital period (the amount of time the two stars take to orbit one another) of a binary system is observed, then the masses of the stars can be calculated using Kepler's third law and Newton's law of gravitation. This makes finding and observing binary systems of great importance to astronomers.***

In the special case of eclipsing binaries, we also have the one direct way for determining the diameters of the stars. By studying the way in which the light dims and brightens, astronomers can measure the sizes of the two stars. There are a few other ways that astronomers measure the sizes of stars, but they are indirect, because those methods require knowing additional information not related to size. For example, one indirect method for calculating the diameter of a star is to derive it from the star's luminosity and surface temperature using the Stefan–Boltzmann equation. Recall from the use of this equation while discussing the Sun that the luminosity is proportional to the star's surface area and the fourth power of its surface temperature. If we know the luminosity and surface temperature (from other observations and calculations) then we can calculate the surface area and determine the diameter.

The Mass and Luminosity Relationship

Once astronomers had determined the masses of a number of stars, they began comparing their masses to their luminosities. The luminosity of a star is the total amount of energy being radiated each second, and it can be calculated based on knowing a star's absolute magnitude. Luminosity is often expressed in comparison to the Sun. A luminosity of "1" means a star has the same luminosity as the Sun, while a luminosity of 100 would mean the star is 100 times more luminous than the Sun. ***Perhaps not surprisingly, it was found that for stars on the main sequence of the HR diagram, mass and luminosity are proportional. The cool, dim, red dwarfs are low-mass stars, while hotter and more luminous blue stars are of higher mass. Yellow stars, like our Sun, lie in between. This proportionality is called the *Mass–Luminosity relationship*.***

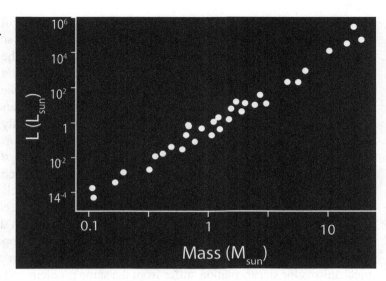

Figure 8.24 The Mass–Luminosity relationship.

8.11 Theories of Stellar Evolution

****Many people mistakenly think that evolution is only about biology. However, evolutionary thinking has permeated all of human endeavors. Evolutionary thinking can be divided into three broad categories: cosmological, geological, and biological. Cosmological evolution concerns itself with the naturalistic origin of the universe and the naturalistic origin of stars.

Astronomers began to develop the modern theories of stellar evolution in the 1950s. According to these theories, stars normally get their energy from the same source that the Sun does—the fusion of hydrogen into helium deep inside or near their cores. Since stars have a finite amount of hydrogen, stars can shine by this mechanism only for a finite length of time. The more mass that a star has, the more hydrogen fuel it has. However, more massive stars are far brighter than less massive stars, so stars that are more massive consume energy far more quickly than lower mass stars. The computed maximum lifetime of the Sun is about 10 billion years. Stars more massive than the Sun have even shorter lifetimes, but less massive stars have longer lifetimes. For instance, the most massive stars could last only a few million years. Most astronomers think that the universe is 13.8 billion years old, so the stars that are more massive than the Sun could not have been around since the beginning of the universe. The least massive stars have computed lifetimes far longer than 13.8 billion years, so some of those stars could date back to the beginning of the universe. Contrast this to the biblical creation model. According to the Bible, God made the stars on Day Four, and that was only thousands of years ago. In this view, all stars have lifetimes far longer than the age of the universe, so all stars date back to the beginning of creation.

How do evolutionary astronomers explain the existence of stars with lifetimes far less than the supposed 13.8-billion-year age of the universe? They hypothesize that most stars formed long after the universe began and that star formation in an ongoing process today. They suggest that the many clouds of gas in space may hold the answer. These clouds (nebulae) are very thin–their density is far less than the best vacuum on Earth. However, the clouds are so large that they often contain thousands of times the mass of the Sun. Gas clouds have roughly the same composition that the Sun and other stars have–they are mostly hydrogen and helium, with a few percent of the remaining elements. Astronomers think that the gravity of these gas clouds help contract the clouds to form stars.

However, there is a problem with this theory. Gas clouds cannot spontaneously contract under their own gravity. While the gas clouds have feeble gravity that slightly pulls the gas inward, the clouds also have pressure that tends to push the gas apart. These forces normally balance so that gas clouds are stable against collapse. To avoid this problem, most astronomers think that an outside force jump-starts the process of collapse. Astronomers have suggested several mechanisms that could do this. One popular idea is that a supernova might explode nearby. As the shock wave of debris from the supernova passes by a cloud, the shock wave could collapse a portion of the cloud, leading to star formation. However, what is a supernova? A supernova is a large explosion that a star might undergo. Where did the star that exploded come from? If that star formed when an earlier supernova exploded, then where did the earlier star that exploded come from? This amounts to a chicken or egg sort of problem. All suggested mechanisms for star formation suffer from the same problem and require that stars first exist, so man's theories here do not really tell us where stars came from. Many astronomers now think that there was a burst of star formation in the early universe, but there is no known mechanism to make this happen.

Stellar evolution

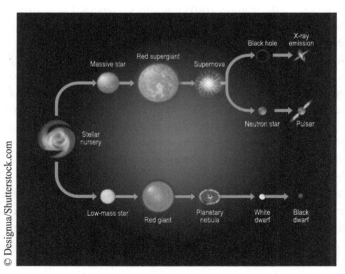

Figure 8.25 A simplified diagram of stellar evolution, with lower-mass stars one day becoming white dwarfs, higher-mass stars exploding and leaving a remnant called a neutron star, and the highest-mass stars exploding and leaving a remnant called a black hole.

From time to time, news reports announce that astronomers have discovered the birth of a star. These stories are reported as if astronomers now see a star that was not there before. However, that cannot be, because the theory of star formation requires that the process takes a very long time, far longer than many human lifetimes. What really ought to be reported is that astronomers have found an object that they think is in some stage of star formation. Astronomers have developed and then modified their theories of star formation over the years as new data became available. Keep in mind that these newly discovered objects amount to snapshots that astronomers arrange within their theories.****

****Laying aside this objection, let us assume that stars do form somehow—what next? Astronomers think that stars end up on the main sequence, where they will remain for their computed lifetimes. For instance, the theory suggests that it took the Sun about 30 million years to reach where it is on the main sequence. The Sun could remain on the main sequence for 10 billion years, and if the Sun is 4.6 billion years, it has exhausted about half of its lifetime on the main sequence. What will happen to the Sun after the 10 billion years have elapsed? After 10 billion years, the Sun will have exhausted the store of hydrogen in its core, so it will look for some new source of energy. Theory suggests that the new source of energy will be fusion of hydrogen into helium in a thin shell around the core. This will be accompanied by a rearrangement of the structure of the Sun so that the Sun gradually will swell in size and cool. The Sun will become a red giant star. Astronomers think that most stars eventually will do this, so this theory explains what red giant stars are—they are evolved stars that once were on the main sequence.

According to the theory, many stars, including the Sun, will tap a few additional sources of energy, but those sources will be short-lived compared to the time spent on the main sequence. Red giant stars are very windy. That is, they shed matter into space. The Sun does this now, in the form of the solar wind. However, the solar wind is extremely feeble compared to the amount of gas blown off red giant stars. Some measured rates of mass loss from red giants would totally evaporate the stars in just thousands of years. Eventually, the mass loss is so great that the core of the star is exposed. The structure of the core of a red giant resembles the structure of a white dwarf, so most astronomers think that this is where white dwarfs come from. Astronomers think that the blown-off gas collects into a complex expanding shell around the newly formed white dwarf. We call these objects **planetary nebulae**. A planetary nebula is a shell of glowing gas. The name is most unfortunate, because the name suggests a connection to planets. There is a superficial resemblance in that through a telescope a planetary nebula appears as a small disk, similar to the appearance of planets. White dwarfs can last for billions of years, but planetary nebulae expand into space and disappear over tens of thousands of years. Planetary nebulae have hot white dwarfs at their centers, but most white dwarfs do not have planetary nebulae around them.****

****Astronomers think that most stars, including the Sun, eventually will end up as white dwarfs. However, according to the theory, the most massive stars have a different fate. Very massive stars are rare. Like the Sun, massive stars get their energy from hydrogen fusion in their cores while on the main sequence. After leaving the main sequence to become giant stars, massive stars probably go through various other sources of energy that the Sun will not. Eventually, the cores of massive stars catastrophically collapse into compact objects. The collapse of the core of a massive star would release a tremendous amount of gravitational potential energy—about as much energy as the Sun would produce in

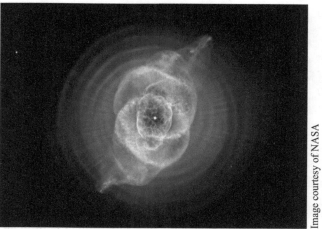

Image courtesy of NASA

Figure 8.26 The Cat's Eye Nebula (NGC 6543) is one of many beautiful planetary nebulae.

Courtesy NASA

Figure 8.27 The bright star (lower left) is the supernova 1994D, a type Ia supernova visible in the galaxy NGC 4526 in the year 1994. The galaxy is 50 to 60 million light years away.

10 billion years. The difference is that this energy is released almost instantly as opposed to over 10 billion years. This energy released so suddenly disrupts the remainder of the star to form a **supernova**.****

****This is just one type of supernova–there are several other types. The types are described by their appearance and spectra, and astronomers explain the different types with their evolutionary theories. A supernova rises to peak brightness very rapidly, over just a few days, and then it more gradually declines in brightness. At peak brightness, supernovae are very bright–they can be brighter than millions of bright stars. Being so bright, we can see them at great distances, even in other galaxies. One type of supernovae, the type Ia, appear to have the same brightness, so they are valuable tools in measuring distances to the galaxies in which they appear. This is an important topic of research in astronomy.****

8.12 Neutron Stars, Black Holes, and Clusters
Neutron Stars

****As discussed above, the collapse of the core of a massive star that leads to a supernova leaves behind a **compact object**. What is a compact object? A compact object is a very massive, but very small, object. Given their large mass and small size, compacts have density so high that we must use modern physics to understand them. We recognize two types of compact objects: neutron stars and black holes. A **neutron star** is a star that consists of just neutrons. The neutrons are packed to the density that exists in the nucleus of an atom. Just one teaspoon of a neutron star on Earth would weigh more than all the automobiles ever manufactured! Neutron stars are only a few miles across, so they are extremely faint. When first predicted in the 1930s, astronomers realized that even if a neutron star were as close as the nearest stars, it would be too faint for us to see, even with the largest telescopes.

However, astronomers failed to appreciate one thing. Stars usually rotate, and so their cores must rotate too. When a spinning object contracts, its rotation speeds up. Think of a spinning ice skater as the skater pulls her arms inward. If the spinning core of a star collapses, the core must spin very rapidly. Stars have magnetic fields too. As a stellar core collapses, the strength of its magnetic field increases tremendously. Usually, the magnetic fields of astronomical bodies do not align with the rotation axes. As bodies rotate, their magnetic fields rotate too. Near a neutron star, there will be charged particles. As the neutron star spins, the magnetic field moves with it. The rapid motion of the strong magnetic field with respect to the charged particles causes the particles to emit radiation. This radiation beams along the poles of the magnetic field. Since the magnetic field spins with the star, the beam of radiation sweeps out a cone shape. If the solar system lies near the cone, it emits a flash of radiation, sort of like a search light beam. The period, or length of time between flashes, is the rotation period of the neutron star. In general, the solar system will not lie along the cone swept out by a pulsar's beam, so most neutron stars cannot be found this way.****

****Jocelyn Bell first found neutron stars while conducting a radio survey of the sky in 1967. In studying point radio sources, Bell found periodic flashes, or pulses, of radio emission coming from some of the radio sources, so astronomers called them **pulsars**. A pulsar is a rapidly rotating neutron star that is oriented so that its periodic flashes of radiation are noticeable to observers on Earth. At first, astronomers did not know what to make of pulsars. Pulsars flashed radiation at very precise periods, about as precise as the best clocks available. For a while, astronomers even considered the possibility that they had intercepted an alien transmission. Astronomers dubbed this possibility LGM, for little green men. Eventually, astronomers figured out that a rapidly rotating neutron star made more sense. Since 1967, astronomers have discovered many more pulsars. There now are thousands of known pulsars.

Pulsars get their power from the rotational kinetic energy of the neutron star. In this way, a spinning neutron star acts as a flywheel on an engine. As a neutron star ages, its rotation ought to slow. The fastest spinning pulsars have periods of about a millisecond (a millisecond is one-thousandth of a second). The longest pulsar periods are a few seconds. Astronomers can measure the rate at which the period is changing, and from that, they can estimate the age. The pulsar in the Crab Nebula is the most famous pulsar. The Crab Nebula is a cloud of glowing gas that on photographs has a faint resemblance to a crab. The Crab Nebula's position corresponds with the position of a supernova that Chinese astronomers observed in the year 1054. Astronomers have determined

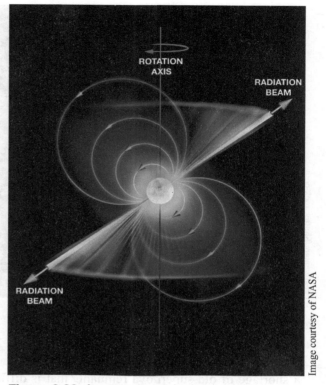

Figure 8.28 A neutron star (pulsar) emits radiation in bright narrow beams that sweep the sky like a lighthouse as the star spins.

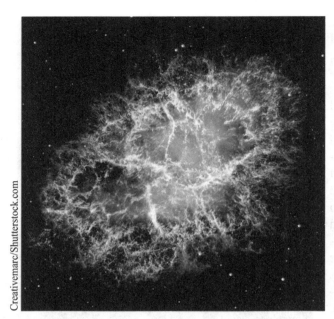

Creativemarc/Shutterstock.com

Figure 8.29 The Crab Nebula, a supernova remnant. One of the first pulsars discovered is inside the Crab Nebula.

the age of the Crab Nebula by measuring its expansion rate, an age that is consistent with the known age of the supernova. Furthermore, astronomers can estimate the age of the pulsar by the pulsar's spin-down rate. That age is consistent with the other two ages.****

****Astronomers consider the Crab Nebula to be a **supernova remnant**. A supernova remnant is the expanding gas blown off the star when it exploded as a supernova. Supernova remnants get thinner as they expand, so eventually they disappear. However, a pulsar will last for a very long time. There are many other known supernova remnants, but because of their relatively short lifetimes, astronomers doubt that we see most of the supernova remnants the supernovae have produced. On the other hand, astronomers are confident that they understand how supernova remnants dissipate over time. There may be a shortage of old supernova remnants that is difficult to explain in a universe that is billions of years old, but not if the universe is only thousands of years old.****

Black Holes

****There is one other type of a compact object–a **black hole**. What is a black hole? Escape velocity is the speed that an object must have at the surface of a massive body, such as a star, to escape that body's gravity. For a given mass and size, we can use physics to calculate the escape velocity of any object. The escape velocity of a black hole exceeds the speed of light, so that not even light can escape from it. Since it gives off no light, a black hole is black. Since material objects cannot travel as fast as light, anything that falls into a black hole cannot come back out. Thus, a black hole truly is a hole in the sense that once things fall in, they cannot come out. Astronomers first suggested that black holes could exist in the eighteenth century, but no one took them seriously until the 1960s, when the term black hole was coined. There is a maximum mass that a neutron star can have, about three times the mass of the Sun. Astronomers think that if a supernova leaves behind a compact object more massive than this, then the compact object must be a black hole.

A black hole is a strange object. The size of a black hole is defined by the size of its event horizon. The *event horizon* is the sphere surrounding a black hole where the escape velocity is equal to the speed of light. Anything inside the event horizon is inside the black hole, and anything outside the event horizon is outside the black hole. What is going on inside the black hole is a mystery, though scientists can observe what is going on outside of it.

How can scientists "see" something that gives off no light? Suppose that a black hole is a member of a close binary star. If the stars orbit one another closely enough, the strong tidal force of the black hole can pull matter off its companion star and on to itself. Because of angular momentum, the matter cannot fall directly onto the black hole, but rather the matter collects in a disk orbiting the black hole. Astronomers call this disk an accretion disk. Astronomers have detected accretion disks around one of the stars in many close binary star systems, so this is well-understood physics. From the inner portion of the accretion disk, matter slowly falls onto the black hole. As the matter falls toward the black hole, it heats up tremendously, and the gas gets so hot that it gives off x-rays. It is not easy to produce large amounts of x-rays, so these x-ray sources stand out. In addition, it is easy to establish whether the x-rays are coming from a close binary star. Astronomers have discovered many x-ray binaries.****

****When astronomers find an x-ray binary, have they necessarily found a black hole? No. Any compact object in a close binary system can produce x-rays, so the compact object could be a neutron star. How is a neutron star different from a black hole? Recall that scientists can determine the masses of the star in a binary star system,

Figure 8.30 An artist's conception of a black hole. The black hole is not visible, but its location is the center of the accretion disk. Notice material streaming in from the lower left, probably from the other star (out of the view here) in the x-ray binary system. Notice two jets of material perpendicular to the accretion disk coming from near the black hole.

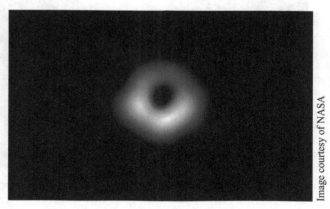

Figure 8.31 This image from 2019 was the first of its kind, capturing light from the accretion disk of hot gases near the event horizon of this black hole.

and that there is an upper limit that the mass of a neutron star can have. If the mass of the compact object exceeds the upper limit of a neutron star, astronomers assume that the compact object must be a black hole. Since the 1960s, astronomers have found many black holes this way.

The story of neutron star and black hole discovery has been filled with surprises. It tells us that the creation is far more interesting that we can imagine.****

© Valerio Pardi/Shutterstock.com

Figure 8.32 The Pleiades.

© Wolfgang Kloehr/Shutterstock.com

Figure 8.33 Globular star cluster M13 found in Hercules.

Star Clusters

****Many stars are members of binary star systems. Other stars, like the Sun, appear to be alone (except for any planets that might orbit them). However, many stars are in **star clusters**. A star cluster is a collection of stars held together by their mutual gravity. There are thousands of star clusters, but most are too faint to see with the naked eye. The brightest appearing star cluster is the Pleiades, or Seven Sisters. The Pleiades are visible each night in the winter and are close to Orion. In fact, the Pleiades are mentioned all three times that Orion is mentioned in the Bible. The Pleiades have a distinctive shape. This shape is recreated in the six stars in the Subaru symbol; Subaru is the Japanese name for the Pleiades.****

****The Pleiades is an example of an **open star cluster**. An open star cluster has an irregular shape and contains a few dozen to a few thousand stars. On the other hand, a **globular star cluster** is spherically shaped and may contain from 50,000 to a million stars. Globular star clusters are less common than open star clusters—Earth's galaxy contains about 200 globular clusters. The nearest globular cluster is thousands of light years away, so the brightest appearing globular clusters are barely visible to the naked eye. However, through a moderate-sized telescope globular star clusters can be very beautiful.****

JOHN WEST: Science & Culture

Dr. John West is Vice President and a Senior Fellow at the Seattle-based Discovery Institute, where he also serves as Associate Director of the Institute's Center for Science & Culture, which he co-founded with philosopher of science Stephen Meyer in 1996. His current research examines the impact of science and "scientism" on public policy and culture. He has written a book called *The Magician's Twin* in which he uses the writings of C.S. Lewis to discuss scientism (when science becomes a religion) and scientocracy (when science is used for political power). Given the events of 2020, this is very timely. What C.S. Lewis wrote about science seventy years ago has been playing out right in front of us. Take notice. What follows is from *The Magician's Twin*.

The Magician's Twin

John G. West

In his classic book *The Abolition of Man* (1944), C.S. Lewis wrote that "[t]he serious magical endeavour and the serious scientific endeavour are twins."[i]

At first reading, Lewis's observation might seem rather strange. After all, science is supposed to be the realm of the rational, the skeptical, and the objective.

Magic, on the other hand, is supposed to be the domain of the dogmatic, the credulous, and the superstitious. Think of a witch doctor holding sway over a tribe of cannibals deep in a South Sea jungle.

As strange as Lewis's observation might first appear, the comparison between science and magic runs throughout a number of his works. The sinister Uncle Andrew in Lewis's Narnian tale *The Magician's Nephew* is both a magician *and* a scientist; and the bureaucratic conspirators at the National Institute of Co-ordinated Experiments (NICE) in Lewis's adult novel *That Hideous Strength* crave the powers of both science and the magician Merlin in their plot to reengineer society.[ii]

For all of the obvious differences between science and magic, Lewis correctly understood that there are at least three important ways in which they are alarmingly similar. More than that, he recognized that these similarities pose a growing threat to the future of civilization as we know it.

---

1. Science as Religion

The first way science and magic are similar according to Lewis is their ability to function as an alternative religion. A magical view of reality can inspire wonder, mystery, and awe. It can speak to our yearning for something beyond the daily activities of ordinary life. Even in our technocratic age, the allure of magic in providing meaning to life can be seen in the continuing popularity of *Star Wars*, *The Lord of the Rings*, the Narnian chronicles, and the adventures of Harry Potter. While magical stories tantalize religious and irreligious people alike, for those without conventional religious attachments, they can provide a substitute spiritual reality.

Modern science can offer a similarly powerful alternative to traditional religion. In Lewis's lifetime, the promoter *par excellence* of this sort of science as religion was popular writer H.G. Wells. Wells and others fashioned Darwin's theory of evolution into a cosmic creation story Lewis variously called "The Scientific Outlook," "Evolutionism," "the myth of evolutionism," and even "Wellsianity."[iii] While some contemporary evolutionists contend that people doubt Darwinian theory because it does not tell a good story,[iv] Lewis begged to differ. In his view, cosmic evolutionism of the sort propounded by Wells was a dramatic narrative brimful of heroism, pathos, and tragedy.

In a bleak and uncaring universe, the hero (life) magically appears by chance on an insignificant planet against astronomical odds. "Everything seems to be against the infant hero of our drama," commented Lewis, " . . . just as everything seems against the youngest son or ill-used stepdaughter at the opening of a fairy tale." No matter, "life somehow wins through. With infinite suffering, against all but insuperable obstacles, it spreads, it breeds, it complicates itself, from the amoeba up to the plant, up to the reptile, up to the mammal."[v] In the words of H.G. Wells, "[a]ge by age through gulfs of time at which imagination reels, life has been growing from a mere stirring in the intertidal slime towards freedom, power and consciousness."[vi]

Through the epic struggle of survival of the fittest, Man himself finally claws his way to the top of the animal kingdom. Eventually he finds Godhood within his grasp if only he will seize the moment. To quote Wells again: "Man is still only adolescent . . . we are hardly in the earliest dawn of human greatness . . . What man has done, the little triumphs of his present state, and all this history we have told, form but the prelude to the things that man has got to do."[vii]

But then, after Man's moment of triumph, tragedy strikes. The Sun gradually cools, and life on Earth is obliterated. In Wells's *The Time Machine*, the protagonist reports his vision of the dying Earth millions of years hence: "The darkness grew apace; a cold wind began to blow . . . From the edge of the sea came a ripple and whisper. Beyond these lifeless sounds the world was silent . . . All the sounds of man, the bleating of sheep, the cries of birds, the hum of insects, the stir that makes the background of our lives—all that was over."[viii]

Lewis explained that he "grew up believing in this Myth and I have felt—I still feel—it's almost perfect grandeur. Let no one say we are an unimaginative age: neither the Greeks nor the Norsemen ever invented a better story." Even now, Lewis added, "in certain moods, I could almost find it in my heart to wish that it was not mythical, but true."[ix]

Lewis did not claim that modern science necessitated the kind of blind cosmic evolutionism promoted by H.G. Wells and company. Indeed, in his book *Miracles* he argued that the birth of modern science and its belief in the regularity of nature depended on the Judeo-Christian view of

God as Creator: "Men became scientific because they expected Law in Nature, and they expected Law in Nature because they believed in a Legislator."[x] Nevertheless, Lewis thought that biology post-Darwin provided potent fuel for turning science into a secular religion.

One does not need to look very far to see science being used in the same way today. In 2012 thousands of atheists and agnostics converged on Washington D.C. for what they called a "Reason Rally."[xi] The rally had all the trappings of an evangelistic crusade, but instead of being preached at by a Billy Graham or a Billy Sunday, attendees got to hear Darwinian biologist Richard Dawkins and *Scientific American* columnist Michael Shermer. Former Oxford University professor Dawkins is known for claiming that "Darwin made it possible to be an intellectually fulfilled atheist," while Shermer once wrote an article "Science is My Savior," which explained how science helped free him from "the stultifying dogma of a 2,000-year-old religion."[xii]

The central role Darwinian evolution continues to play for the science-as-religion crowd is readily apparent in the countless "Darwin Day" celebrations held around the globe each year on February 12, Charles Darwin's birthday. Darwin Day is promoted by a group calling itself the International Darwin Day Foundation. Managed by the American Humanist Association, the group's mission is "to encourage the celebration of Science and Humanity" because "[s]cience is our most reliable knowledge system."[xiii]

According to Amanda Chesworth, one of the co-founders of the Darwin Day movement, the purpose of Darwin Day is to "recognize and pay homage to the indomitable minds and hearts of the people who have helped build the secular cathedrals of verifiable knowledge." Chesworth's word choice is particularly astute: By doing science, scientists in her view are building "secular *cathedrals*."[xiv] The iconography of religion is unmistakable. In the words of one Darwin Day enthusiast who posted an approving comment on the official Darwin Day site: "To me, Charles Darwin is more of a God than the one armies had nailed to a cross."[xv]

Perhaps the most tireless proponents of cosmic evolutionism today are the husband and wife team of Michael Dowd and Connie Barlow, who bill themselves as "America's Evolutionary Evangelists."[xvi] A former evangelical Christian turned Unitarian minister turned religious "naturalist," Dowd is author of *Thank God for Evolution!*, the subtitle of which is "How the Marriage of Science and Religion Will Transform Your Life and Our World."[xvii] Dowd calls his brand of cosmic evolutionism the "Great Story," which is defined on the Great Story website as "humanity's **sacred narrative of an evolving Universe** of emergent complexity and breathtaking creativity—a story that offers each of us the opportunity to find meaning and purpose in our lives and our time in history."[xviii] The Great Story comes along with its own rituals, parables, hymns, sacred sites, "evolutionary revival" meetings, Sunday School curricula, and even "cosmic rosaries," necklaces of sacred beads to teach children the fundamental doctrines of cosmic evolutionism.[xix]

Dowd has attracted widespread support from Nobel laureates, atheistic evolutionists, and theistic evolutionists. For all of his outreach to the faith community, however, Dowd dismisses the reality of God just as much as atheist biologist Richard Dawkins. In an article written for *Skeptic* magazine, Dowd acknowledged his view that "God" is simply a myth: "God is not a person; God is a personification of one or more deeply significant dimensions of reality."[xx] Just as people in the ancient world personified the oceans as the god Poseidon or the Sun as the god Sol, contemporary people personify natural forces and call them "God."[xxi] Hence, Dowd's Great Story is ultimately a drama of the triumph of blind and undirected matter in a universe where a Creator does not actually exist. This becomes explicit in the description of the Great Story provided by philosopher Loyal Rue cited approvingly on the Great Story website:

> In the course of epic events, matter was distilled out of radiant energy, segregated into galaxies, collapsed into stars, fused into atoms, swirled into planets, spliced into molecules, captured into cells, mutated into species, compromised into thought, and cajoled into cultures. **All of this (and much more)** is what **matter** has done as systems upon systems of organization have emerged over thirteen billion years of creative natural history.[xxii]

"All of this . . . is what *matter* has done," *not* God. Just like the narrative promoted by H.G. Wells and the scientific materialists at the beginning of the twentieth century, the cosmic evolutionism offered by Dowd and his followers in the twenty-first century is ultimately reducible to scientific materialism. The bottom line of their secular creation story is neatly encapsulated by Philip Johnson: "In the beginning were the particles. And the particles somehow became complex living stuff. And the stuff imagined God, but then discovered evolution."[xxiii]

Lewis would not have been surprised by current efforts to co-opt traditional religion in the name of science, or even to find a lapsed clergyman leading the charge. In Lewis's novel *That Hideous Strength*, the sometime clergyman Straik joins hand-in-hand with the avowed scientific materialists in the name of promoting a new this-worldly religion. As the impassioned Rev. Straik declares to Mark Studdock: "The Kingdom is going to arrive: in this world: in this country. The powers of science are an instrument. An irresistible instrument."[xxiv]

2. Science as Credulity

The second way science and magic are similar according to Lewis is their encouragement of a stunning *lack* of skepticism. This may seem counterintuitive, since science in the popular imagination is supposed to be based on logic and evidence, while magic is supposed to be based on a superstitious acceptance of claims made in the name of the supernatural. In the words of Richard Dawkins, "Science is based upon verifiable evidence," while "[r]eligious faith" (which Dawkins views as a kind of magic) "not only lacks evidence, its independence from evidence is its pride and joy."[xxv] Yet as Lewis well knew, scientific thinking no less than magical thinking can spawn a kind of credulity that accepts every kind of explanation no matter how poorly grounded by the facts. In the age of magic, the claims of the witch-doctor were accepted without contradiction. In the age of science, almost anything can be taken seriously if only it is defended in the name of science.

Lewis explained that one of the things he learned by giving talks at Royal Air Force camps during World War II was that the "real religion" of many ordinary Englishmen was a completely uncritical "faith in 'science.'"[xxvi] Indeed, he was struck by how many of the men in his audiences "did not really believe that we have any reliable knowledge of historic man. But this was often curiously combined with a conviction that we knew a great deal about Pre-Historic Man: doubtless because Pre-Historic Man is labelled 'Science' (which is reliable) whereas Napoleon or Julius Caesar is labelled as 'History' (which is not)."[xxvii]

But it was not just the "English Proletariat" who exhibited a credulous acceptance of claims made in the name of science according to Lewis. In *That Hideous Strength*, when the young sociologist Mark Studdock expresses doubts that NICE can effectively propagandize "educated people," the head of NICE's police force, Fairy Hardcastle, responds tartly: "Why you fool, it's the educated reader who *can* be gulled. All our difficulty comes with the others. When did you meet a workman who believes the papers? He takes it for granted that they're all propaganda and skips the leading articles . . . We have to recondition him. But the educated public, the people who read the highbrow weeklies, don't need reconditioning. They're all right already. They'll believe anything."[xxviii]

For Lewis, two leading examples of scientism-fueled gullibility of the intellectual classes during his own day were Freudianism and evolutionism.

Lewis's interest in Freud dated back to his days as a college student. In his *Surprised by Joy* (1955), he recalled how as an undergraduate "the new Psychology was at that time sweeping through us all. We did not swallow it whole . . . but we were all influenced."[xxix] In 1922, he recorded in his diary a discussion with friends saying that "[w]e talked a little of psychoanalysis, condemning Freud."[xxx] Although skeptical of Freud, Lewis remained intrigued, for a just few weeks later he notes that he was reading Freud's *Introductory Letters on Psychoanalysis.*[xxxi]

A decade later, and shortly after Lewis had become a Christian, Freud made a cameo appearance in *The Pilgrim's Regress* (1933), Lewis's autobiographical allegory of his intellectual and spiritual journey toward Christianity.[xxxii] In Lewis's story, the main character John ends up being arrested and flung into a dungeon by a stand-in for Freud named Sigismund Enlightenment (Sigismund was Sigmund Freud's full first name).[xxxiii] The dungeon is overseen by a Giant known as the Spirit of the Age who makes people transparent just by looking at them. As a result, wherever John turns, he sees through his fellow prisoners into their insides. Looking at a woman, he sees through her skull and "into the passages of the nose, and the larynx, and the saliva moving in the glands and the blood in the veins: and lower down the lungs panting like sponges, and the liver, and the intestines like a coil of snakes."[xxxiv] Looking at an old man, John sees the man's cancer growing inside him. And when John turns his head toward himself, he is horrified to observe the inner workings of his own body. After many days of such torment, John cries out in despair: "I am mad. I am dead. I am in hell for ever."[xxxv]

The dungeon is the hell of materialistic reductionism, the attempt to reduce every human trait to an irrational basis, all in the name of modern science. Lewis saw Freud as one of the trailblazers of the reductionist approach. By attempting to uncover the "real" causes of people's religious and cultural beliefs in their subconscious and irrational urges and complexes, Freud eroded not only their humanity, but the authority of rational thought itself.

In the 1940s, Lewis offered an explicit critique of Freudianism in a lecture to the Socratic Society at Oxford. Noting that people used to believe that "if a thing seemed obviously true to a hundred men, then it was probably true in fact," Lewis observed that "[n]owadays the Freudian will tell you to go and analyze the hundred: you will find that they all think Elizabeth [I] a great queen because they all have a mother-complex. Their thoughts are psychologically tainted at the source."[xxxvi]

"Now this is obviously great fun," commented Lewis, "but it has not always been noticed that there is a bill to pay for it." If all beliefs are thus tainted at the source and so should be disregarded, then what about Freud's own system of belief? The Freudians "are in the same boat with all the rest of us . . . They have sawn off the branch they were sitting on."[xxxvii] In the name of a scientific study of psychology, the Freudians had undercut the confidence in reason needed for science itself to continue to flourish.[xxxviii]

Evolutionism was another prime example of credulous thinking fostered by scientism according to Lewis. As Chapter 6 will explain, Lewis did not object in principle to an evolutionary process of common descent, although he was skeptical in practice of certain claims about common descent. But Lewis had no patience for the broader evolutionary idea that matter magically turned itself into complex and conscious living things through a blind and undirected process. Lewis lamented that "[t]he modern mind accepts as a formula for the universe in general the principle 'Almost nothing may be expected to turn into almost everything' without noticing that the parts of the universe under our direct observation tell a quite different story."[xxxix] Fueled by "Darwinianism," this sort of credulity drew on "a number of false analogies" according to Lewis: "the oak coming from the acorn, the man from the spermatozoon, the modern steamship from the primitive coracle. The supplementary truth that every acorn was dropped by an oak, every spermatozoon derived from a man, and the first boat by something so much more complex than itself as a man of genius, is simply ignored."[xl]

Lewis also thought that evolutionism, like Freudianism, promoted a "fatal self-contradiction" regarding the human mind.[xli] According to the Darwinian view, "reason is simply the unforeseen and unintended by-product of a mindless process at one stage of its endless and aimless becoming." Lewis pointed out the fundamental difficulty with this claim: "If my own mind is a product of the irrational—if what seem my clearest reasonings are only the way in which a creature conditioned as I am is bound to feel—how shall I trust my mind when it tells me about Evolution?" He added that "[t]he fact that some people of scientific education cannot by any effort be taught to see the difficulty, confirms one's suspicion that we here touch a radical disease in their whole style of thought."[xlii]

Although science is supposed to be based on logic, evidence, and critical inquiry, Lewis understood that it could be easily misused to promote uncritical dogmatism, and he lived during an era in which this kind of misuse of science was rampant. Consider the burgeoning "science" of eugenics, the effort to breed better human beings by applying Darwinian principles of selection through imprisonment, forced sterilization, immigration restrictions, and other methods. Generally regarded today as pseudoscience, eugenics originated with noted British scientist Francis Galton (Charles Darwin's cousin), and it found widespread popularity in Lewis's day among elites in England, the United States, and Germany. Eugenics was the consensus view of the scientific community during much of Lewis's lifetime, and those who opposed it were derided as anti-science reactionaries or religious zealots standing in the way of progress. In America, its champions included members of the National Academy of Sciences and evolutionary biologists at the nation's top research universities.[xliii] In Britain, noted eugenists included evolutionary biologist Julian Huxley, grandson of "Darwin's Bulldog" Thomas Henry Huxley. Julian Huxley complained that in civilized societies "the elimination of defect by natural selection is largely . . . rendered inoperative by medicine, charity, and the social services." As a result, "[h]umanity will gradually destroy itself from within, will decay in its very core and essence, if this slow but relentless process is not checked."[xliv]

The United States holds the dubious honor of enacting the world's first compulsory eugenics sterilization law, but it was Nazi Germany that pursued eugenics with special rigor in the 1930s and

40s. Not content with merely sterilizing hundreds of thousands of the so-called "unfit," Nazi doctors eventually started killing handicapped persons en masse in what turned out to be a practice run for Hitler's extermination campaign against the Jews.[xlv]

The horrors of Nazi eugenics effectively killed off enthusiasm for eugenics in the mainstream scientific community after World War II. But there were other cases where scientific elites showed a similarly breathtaking lack of skepticism during this period. In the field of human evolution, much of the scientific community was hoodwinked for two generations into accepting the infamous Piltdown skull as a genuine "missing link" between humans and their ape-like ancestors before the fossil was definitively exposed as a forgery in 1953 (much to Lewis's private amusement).[xlvi] In the field of medicine, the lobotomy was embraced as a miracle cure by large parts of the medical community well into the 1950s, and the scientist who pioneered the operation in human beings even won a Nobel Prize for his efforts in 1949. Only after tens of thousands of individuals had been lobotomized (including children) did healthy skepticism begin to prevail.[xlvii] And in the field of human sexuality, Darwinian zoologist Alfred Kinsey's studies on human sex practices were accepted uncritically by fellow researchers and social scientists for decades despite the fact that his wildly unrepresentative samples and coercive interview techniques made his research little more than junk science.[xlviii]

If scientists themselves could demonstrate such stunning bouts of credulity about scientific claims, members of the general public were even more susceptible to the disease according to Lewis. In an age of science and technology, Lewis knew that ordinary citizens must increasingly look to scientific experts for answers, and that would likely lead people to defer more and more to the scientists, letting the scientists do their thinking for them and neglecting their own responsibilities for critical thought in the process.

Lewis knew firsthand the dangers of simply deferring to scientific claims, recalling that his own atheistic "rationalism was inevitably based on what I believed to be the findings of the sciences, and those findings, not being a scientist, I had to take on trust—in fact, on authority."[xlix] Lewis understood that the ironic result of a society based on science might be greater credulity, not less, as more people simply accepted scientific claims on the basis of authority. This was already happening in his view. Near the end of his life, Lewis observed that "the ease with which a scientific theory assumes the dignity and rigidity of fact varies inversely with the individual's scientific education," which is why when interacting "with wholly uneducated audiences" he "sometimes found matter which real scientists would regard as highly speculative more firmly believed than many things within our real knowledge."[l] In Lewis's view, the increasing acquiescence of non-scientists to those with scientific and technical expertise gave rise to by far the most dangerous similarity between science and magic, one that threatened the future of Western civilization itself.

3. Science as Power

The third and most significant way science is similar to magic according to Lewis is its quest for power. Magic wasn't just about understanding the world; it was about controlling it. The great wizard or sorcerer sought power over nature. Similarly, science from the beginning was not just the effort to understand nature, but the effort to control it. "For magic and applied science alike the problem is how to subdue reality to the wishes of men," wrote Lewis. In pursuit of that objective, both magicians and scientists "are ready to do things hitherto regarded as disgusting and impious—such as digging up and mutilating the dead."[li]

Of course, there is a critically important difference between science and magic: Science works, while magic is relegated today to the pages of the fairy tale. Science cures diseases. Science increases food production. Science puts men on the Moon and ordinary people in jet planes. Science fills our homes with computers, iPhones, and microwave ovens. Herein lies the great temptation of modern science to modern man. The world as we know it faces apparently insurmountable evils from hunger to disease to crime to war to ecological devastation. Science offers the hope of earthly salvation through the limitless creativity of human ingenuity—or so the prophets of scientism have claimed over the past century, including H.G. Wells and evolutionary biologists J.B.S. Haldane and Julian Huxley during C.S. Lewis's own day. Haldane viewed science as "man's gradual conquest, first of space and time, then of matter as such, then of his own body and those of other living beings, and finally the subjugation of the dark and evil elements in his own soul,"[lii] and he urged his fellow scientists to no longer be "passively involved in the torrent of contemporary history, but actively engaged in changing society and shaping the world's future."[liii]

C.S. Lewis was not persuaded. In his view, the scientific utopians failed to take into account the moral vacuum at the heart of contemporary science. Lewis stressed that he was not anti-science; but he still worried that modern science was ill-founded from the start: "It might be going too far to say that the modern scientific movement was tainted from its birth: but I think it would be true to say that it was born in an unhealthy neighbourhood and at an inauspicious hour."[liv] Lewis noted that modern science attempts to conquer nature by demystifying its parts and reducing them to material formulas by which they can be controlled. The results of this materialistic reductionism are often laudable (e.g., antibiotics, personal computers, and the invention of airplanes). Nevertheless, when the conquest of nature is turned on man himself, a problem arises: "[A]s soon as we take the final step of reducing our own species to the level of mere Nature, the whole process is stultified, for this time the being who stood to gain and the being who has been sacrificed are one and the same."[lv] By treating human beings as the products of blind non-rational forces, scientific reductionism eliminates man as a rational moral agent. In Lewis's words, "[m]an's final conquest has proved to be the abolition of Man."[lvi]

Lewis worried that scientism's reductionist view of the human person would open the door wide to the scientific manipulation of human beings. "[I]f man chooses to treat himself as raw material," he wrote, "raw material he will be: not raw material to be manipulated, as he fondly imagined, by himself, but by mere appetite, that is, mere Nature, in the person of his dehumanized Conditioners."[lvii] Lewis thought there would be no effective limits on human manipulation in the scientific age because scientism undermined the authority of the very ethical principles needed to justify such limits. According to scientism, old cultural rules (such as "Man has no right to play God" or "punishment should be proportionate to the crime") were simply the by-products of a blind evolutionary process and could be disregarded or superseded as needed. Thus, any restrictions on the application of science to human affairs ultimately would be left to the personal whims of the elites.

Lewis's concern about the powerful impact of scientism on society was detectable already in *Dymer* (1926) and *The Pilgrim's Regress* (1933), but by the late 1930s and early 1940s his alarm was on full display in his science fiction trilogy, which he continued to publish as the world plunged into another world war. It is significant that Lewis spent World War II writing not about the dangers of Nazism or communism (even though he detested both), but about the dangers of scientism and its effort to abolish man.[lviii] Scientism was a greater threat in Lewis's view than fascism or communism because it infected representative democracies like Britain no less than totalitarian societies: "The process which, if not checked, will abolish Man, goes on apace among Communists and Democrats no less than among Fascists." Lewis acknowledged that "[t]he methods may (at first) differ in brutality" between scientism and totalitarianism, but he went on to make a shocking claim: "[M]any a mild-eyed scientist in pince-nez, many a popular dramatist, many an amateur philosopher in our midst, means in the long run just the same as the Nazi rulers of Germany."[lix]

That message lies at the heart of Lewis's novel *That Hideous Strength*, written in 1942 and 1943, but not published until 1945.[lx] As previously mentioned, *That Hideous Strength* tells the story of a sinister conspiracy to turn England into a scientific utopia. The vehicle of transformation is to be a lavishly funded new government bureaucracy with the deceptively innocuous name of the National Institute for Co-ordinated Experiments, or NICE for short.[lxi] Of course, there is nothing nice about NICE. Its totalitarian goal is to meld the methods of modern science with the coercive powers of government in order "to take control of our own destiny" and "make man a really efficient animal." The Institute's all-encompassing agenda reads like a wish list drawn up by the era's leading scientific utopians: "sterilization of the unfit, liquidation of backward races (we don't want any dead weights), selective breeding," and "real education," which means "biochemical conditioning . . . and direct manipulation of the brain."[lxii] NICE's agenda also includes scientific experimentation on both animals and criminals. The animals would be "cut up like paper on the mere chance of some interesting discovery," while the criminals would no longer be punished but cured, even if their "remedial treatment" must continue indefinitely.[lxiii]

Lewis lampoons the scientific bureaucrats running NICE, and he relishes pointing out just how narrow minded and parochial they are for all of their supposed sophistication. This comes out clearly when Mark Studdock and a fellow researcher from the sociology branch of NICE (Cosser) visit a picturesque country village in order to write a report advocating its demolition. Mark, who is not quite as far down the path of scientism as Cosser, feels like he is "on a holiday" while visiting the village, enjoying the natural beauty of the sunny winter day, relaxing at a pub for a drink, and feeling the aesthetic attraction of historic English architecture. Cosser is impervious to such things, placing no value on anything outside his narrow field of sociological expertise. Instead of delighting in the beauty of nature, Cosser complains about the "[b]loody awful noise those birds make."[lxiv] Instead of enjoying a drink at the pub, he complains about the lack of ventilation and suggests that the alcohol could be "administered in a more hygienic way." When Mark suggests that Cosser is missing the point of the pub as a gathering place for food and fellowship, Cosser replies "Don't know, I'm sure . . . Nutrition isn't my subject. You'd want to ask Stock about that." When Mark mentions that the village has "its pleasant side" and that they need to make sure that whatever it is replaced with is something better in all areas, "not merely in efficiency," Cosser again pleads that this is outside his area. "Oh, architecture and all that," he replies. "Well, that's hardly my line, you know. That's more for someone like Wither. Have you nearly finished?"[lxv] A hyper-specialist,

Cosser can't see past his proverbial nose. Yet he is being given the power to decide whether to dispossess members of an entire village from their homes.[lxvi]

That Hideous Strength resonated with the public, and it quickly became Lewis's most popular adult novel, despite negative reviews from critics, including one from J.B.S. Haldane, who thought the novel was a blatant attack on science.[lxvii] It is easy to understand why the public of the 1940s might have been receptive to the novel's message. Two world wars and the rise of totalitarianism in Germany and Russia had dampened popular enthusiasm for the message of the scientific utopians. After all, it was hard to view science as savior when scientists were busy bringing forth poison gas, the V-2 rocket, and the atomic bomb—not to mention new methods of killing the handicapped in the name of eugenics in Germany. To many people, the new age ushered in by science looked more like a nightmare than a paradise.

After World War II, however, even the looming threat of nuclear annihilation did not prevent some from renewing their quest for societal salvation through science, and scientific utopianism began to revive. At the global level, Julian Huxley called for bringing about a better future by promoting "scientific world humanism, global in extent and evolutionary in background,"[lxviii] while in America renewed optimism toward science was exemplified by icons of pop culture such as Walt Disney's "Tomorrowland" in Disneyland, *The Jetsons* cartoon series, and the 1962 World's Fair in Seattle, which celebrated the seemingly endless possibilities of the science-led world of "Century 21."

For his part, Lewis continued to sound the alarm about the dangers of what he variously called "technocracy" or "Scientocracy"—government in the name of science that is disconnected from the traditional limits of both morality and a free society.[lxix] Lewis's most eloquent post-war statement on the subject came in the article "Willing Slaves of the Welfare State" published in *The Observer* in 1958. In that essay, Lewis worried that we were seeing the rise of a "new oligarchy [that] must more and more base its claim to plan us on its claim to knowledge . . . This means they must increasingly rely on the advice of scientists, till in the end the politicians proper become merely the scientists' puppets."[lxx] Lewis believed that the world's "desperate need[s]" of "hunger, sickness, and the dread of war" would make people all too willing to accept an "omnicompetent global technocracy," even if it meant surrendering their freedoms. "Here is a witch-doctor who can save us from the sorcerers—a war-lord who can save us from the barbarians—a Church that can save us from Hell. Give them what they ask, give ourselves to them bound and blindfold, if only they will!"[lxxi]

Lewis did not deny that scientific and technical knowledge might be needed to solve our current problems. But he challenged the claim that scientists had the right to rule merely because of their superior technical expertise. Scientific knowledge may be necessary for good public policy in certain areas, but Lewis knew that it was hardly sufficient. Political problems are preeminently moral problems, and scientists are ill-equipped to function as moralists according to Lewis: "Let scientists tell us about sciences. But government involves questions about the good for man, and justice, and what things are worth having at what price; and on these a scientific training gives a man's opinion no added value." [lxxii]

Lewis's warnings about the threat of scientocracy could have come from the latest headlines. Since the 1990s there has been a dramatic increase in what some have called the "authoritarian tone" of science, exemplified by the growing use in science journalism during this period of phrases such as "science requires," "science dictates," and "science tells us we should."[lxxiii] The changes in journalism track with similar developments in politics and public policy. Whether the topic be embryonic stem cell research, climate change, health insurance mandates, the teaching of evolution, or any number of other topics, "science" is increasingly being used as a trump card in public debates to suppress dissent and curtail discussion. Regardless of the issue, experts assert that their public policy positions are dictated by "science," which means that anyone who disagrees with them is "anti-science."

The conflict over government funding for embryonic stem cell research is a perfect example. Oppose taxpayer funding for embryonic stem cell research, and you are guaranteed to be labeled "anti-science" as well as a religious fanatic. However, this storyline of enlightened scientists versus intolerant fundamentalists opposed to research obscures the complexities of the actual debate. First, there are plenty of scientific (as opposed to ethical or religious) objections to the efficacy of embryonic stem cell research; these are conveniently ignored by framing the dispute as science versus anti-science.[lxxiv] Second, raising ethical questions about certain kinds of scientific research makes one "anti-science" only if one accepts scientism's premise that science is the one valid form of knowledge in the public square and scientific research therefore should operate free from any outside restrictions whatever. According to this mindset, opposition to the infamous Tuskegee syphilis experiments or Nazi medical experimentation on Jews would make one "anti-science." But that is ridiculous. Practicing science does not require operating in a moral vacuum, and raising ethical objections to some forms of scientific research does not make one "anti-science."

A similar situation exists in the debate over climate change. Question any part of the climate change "consensus" (how much climate change is going on, how much humans contribute to it, and what should humans do about it), and one is instantly declared "anti-science" or even a threat to the future of the human race. The goal of this kind of rhetoric is not to win by persuading others, but by silencing them.

Along with the growing use of science as a trump card, we are seeing the revival of scientific justifications for eugenics under the banners of "Transhumanism" (see Chapter 10) and "reprogenetics." The latter term was coined by Princeton University biologist Lee Silver, who urges human beings to take control of their evolution and evolve themselves into higher race of beings with god-like powers.[lxxv] Although Silver is concerned that the supposed blessings of genetic engineering might not be equally distributed across the population,[lxxvi] he nonetheless urges us to seize the opportunity: "[H]uman beings . . . now have the power not only to control but to create new genes for themselves. Why not seize this power? Why not control what has been left to chance in the past?[lxxvii] "Transhumanism" and "reprogenetics" may still sound like science fiction to many people, but eugenic abortions targeting children with genetic defects are already well underway. In 2012, physician Nancy Snyderman, chief medical editor for NBC News, publicly defended eugenic abortions on national television squarely on the basis of science: "I am pro-science, so I believe that this is a great way to prevent diseases."[lxxviii] Of course, if it is "pro-science" to support eradicating babies with genetic flaws, it must be "anti-science" to oppose it.

For the moment, the new eugenics is focused more on encouraging individuals to willingly breed a better race than on imposing top-down measures, but the use of science as a justification for coercion is on the upswing as well:

- In the name of saving the planet from global warming, British scientist James Lovelock has called for the suspension of democracy: "Even the best democracies agree that when a major war approaches, democracy must be put on hold for the time being. I have a feeling that climate change may be an issue as severe as a war. It may be necessary to put democracy on hold for a while."[lxxix]
- In the name of promoting biodiversity, evolutionary zoologist Eric Pianka at the University of Texas urges the reduction of the Earth's human population by up to 90% and calls on the government to confiscate all the earnings of any couple who has more than two children. "You should have to pay more when you have your first kid—you pay more taxes," he insists. "When you have your second kid you pay a lot more taxes, and when you have your third kid you don't get anything back, they take it all."[lxxx]
- In order to achieve the admittedly laudable goal of ending obesity, Harvard evolutionary biologist Daniel Lieberman advocates coercive measures by the government to control our diets. Lieberman argues that coercion is necessary because evolutionary biology shows us that we cannot control our sugar intake on our own power. "We have evolved to need coercion."[lxxxi]
- When the Obama administration mandated that many private religious employers include contraceptives and even certain kinds of abortion drugs as part of their health care plans, the abrogation of religious liberty rights was justified in the name of science. "Scientists have abundant evidence that birth control has significant health benefits for women," declared Secretary of Health and Human Services Kathleen Sebelius, defending the mandate.[lxxxii]

Lewis's age of Scientocracy has come upon us with a vengeance. Now we need to figure out what to do about it.

A Regenerate Science?

Lewis provides a hint as to what will be required to overcome scientism in his Narnian story *The Magician's Nephew*. Despite its title, there are actually two magicians in the story. The first, Uncle Andrew, embodies the longing to fuse both science and magic. Although a magician, Uncle Andrew is also a scientist. He has a microscope, and he experiments on animals.[lxxxiii] By pursuing power over nature without regard to ethics, Uncle Andrew sets in motion a train of events that ultimately brings a far greater magician, Queen Jadis, into both Earth and Narnia, which she thereupon threatens to enslave. Jadis previously destroyed her own world, Charn, after using her knowledge of "the Deplorable Word" to liquidate the entire population of the planet. The "Deplorable Word" was a secret formula "which, if spoken with the proper ceremonies, would destroy all living things except the one who spoke it." Previous rulers of Charn had pledged never to seek knowledge of the formula, but Jadis violated her oath, and when faced with defeat in battle, she decided to use the word.[lxxxiv]

Jadis is ultimately thwarted in her effort to take over new worlds not by the actions of a fellow magician, but by the repentance of a young boy, Digory. Digory's unconstrained curiosity previously had brought Jadis out of a deep sleep. In order to undo the harm brought about by awakening Jadis, Digory promises Aslan, the Creator of Narnia, that he will journey to a garden on top of the mountains where he will pick a magical apple and bring it back to Aslan. When Digory arrives at the garden, he finds Jadis already there, having gorged herself on one of the apples despite a sign forbidding people to take apples for themselves. Jadis then urges Digory to disregard his promise to Aslan and take an apple for his dying mother, assuring him that the apple will heal her of her illness. Even when Jadis accuses Digory of being "heartless" for not being willing to save his own mother, Digory rebuffs the temptation to break faith with Aslan. As a result of Digory's unwillingness to cooperate with her evil scheme, Jadis and her evil power are kept in check for many centuries.[lxxxv]

The Magician's Nephew was written during the 1950s, the very period when Lewis's concerns about an "omnicompetent global technocracy" continued to grow. Jadis clearly represents the dangers of scientism. Her use of the "Deplorable Word" in her own world is perhaps a commentary on the age of nuclear weapons and our own efforts to develop ever more destructive weapons of mass destruction. After Aslan says that humans should take warning from the destruction of Charn, Digory's friend Polly says: "But we're not quite as bad as that world, are we, Aslan?" Aslan responds: "Not yet. But you are growing more like it. It is not certain that some wicked one of your race will not find out a secret as evil as the Deplorable Word and use it to destroy all living things." Aslan then tells Digory and Polly that "before you are an old man and an old woman, great nations in your world will be ruled by tyrants who care no more for joy and justice and mercy than the Empress Jadis. Let your world beware."[lxxxvi] Since *The Magician's Nephew* is set in the early 1900s, Aslan is undoubtedly referring to the two world wars and subsequent "Cold War" that loomed on the horizon, all of which would be accompanied by horrifying new uses of science and technology to kill and manipulate humanity.[lxxxvii]

In *The Abolition of Man*, Lewis expressed his hope that a reformation of science could be brought about by scientists. But he made clear that the task was too important to be left to them alone: "[I]f the scientists themselves cannot arrest this process before it reaches the common Reason and kills that too, then someone else must arrest it."[lxxxviii] In a free society, scientism requires the cooperation of scientists and non-scientists alike to prevail, and it requires the cooperation of both scientists and non-scientists to be defeated.

Like Digory, people today need the courage and independence of thought to stand up to the magicians of scientism. They need to be willing to ask questions, challenge assumptions, and defend a broader view of rationality than that permitted by scientific materialism. Whether the issue is climate change, embryonic stem cell research, genetic engineering, evolution and intelligent design, or something else, it is not enough to simply acquiesce to the current "climate of opinion" in science or anything else, as Lewis himself well knew. "I take a very low view of 'climates of opinion,'" he commented, noting that "[i]n his own subject every man knows that all discoveries are made and all errors corrected by those who *ignore* the 'climate of opinion.'"[lxxxix]

At the end of *The Abolition of Man*, Lewis issued a call for a "regenerate science" that would seek to understand human beings and other living things as they really are, not try to reduce them to automatons. "When it explained it would not explain away. When it spoke of the parts it would remember the whole. While studying the *It* it would not lose what Martin Buber calls the *Thou*-situation."[xc]

Lewis was not quite sure what he was asking for, and he was even less sure that it could come to pass. Yet in recent decades we have begun to see glimmers. New developments in biology, physics, and cognitive science are raising serious doubts about the most fundamental tenets of scientific materialism. In physics, our understanding of matter itself is becoming increasingly non-material.[xci]

In biology, scientists are discovering how irreducibly complex biological systems and information encoded in DNA are pointing to the reality of intelligent design in nature.[xcii] In cognitive science, efforts to reduce mind to the physical processes of the brain continue to fail, and new research is providing evidence that the mind is a non-reducible reality that must be accepted on its own terms.[xciii] What George Gilder has called "the materialist superstition" is being challenged as never before.[xciv]

Nearly 50 years after C.S. Lewis's death, we are facing the possibility that science can become something more than the magician's twin. Even in the face of surging scientism in the public arena, an opportunity has opened to challenge scientism on the basis of science itself, fulfilling Lewis's own desire that "from Science herself the cure might come."[xcv] Let us hope we find the clarity and courage to make the most of the opportunity.

Notes

i.   C.S. Lewis, *The Abolition of Man* (New York: Macmillan, 1955), 87.

ii. C.S. Lewis, *The Magician's Nephew* (New York: Macmillan, 1955), and *That Hideous Strength* (New York: Macmillan, 1965).

iii. C.S. Lewis, "Is Theology Poetry" in *The Weight of Glory and Other Addresses*, revised and expanded edition (New York: Macmillan, 1980), 79; C.S. Lewis, "The Funeral of a Great Myth," in *Christian Reflections*, ed. Walter Hooper (Grand Rapids: Eerdmans, 1967), 82–93; C.S. Lewis, "The World's Last Night," in *The World's Last Night and Other Essays* (San Diego: Harcourt Brace & Company, 1960), 101.

iv. Tom Bartlett, "Is Evolution a Lousy Story," *The Chronicle of Higher Education* (May 2, 2012), accessed June 12, 2012, http://chronicle.com/blogs/percolator/is-evolution-a-lousy-story/29158?sid=pm&utm_source=pm&utm_medium=en.

v.   Lewis, "Is Theology Poetry?" 79.

vi. H.G. Wells, *A Short History of the World* (New York: Macmillan, 1922), 16, accessed June 13, 2012, http://www.gutenberg.org/files/35461/35461-h/35461-h.htm.

vii. Ibid., 426–427.

viii. H.G. Wells, *The Time Machine* (1898), chapter 12, accessed June 13, 2012, http://www.gutenberg.org/cache/epub/35/pg35.html.

ix. Lewis, "Funeral of a Great Myth," 88.

x.   C.S. Lewis, *Miracles: A Preliminary Study* (New York: Macmillan, 1960), 106.

xi. Cathy Lynn Grossman, "Richard Dawkins to Atheist Rally: 'Show Contempt' for Faith," *USA Today,* March 24, 2012, accessed June 12, 2012, http://content.usatoday.com/communities/Religion/post/2012/03/-atheists-richard-dawkins-reason-rally/1#.T8AiFb-iyJg; Kimberly Winston, "Atheists Rally on National Mall," *The Huffington Post*, March 24, 2012, accessed June 12, 2012, http://www.huffingtonpost.com/2012/03/24/atheist-rally_n_1377443.html.

xii. Richard Dawkins, *The Blind Watchmaker: Why the Evidence of Evolution Reveals a Universe Without Design* (New York: W.W. Norton and Co., 1996), 6; Michael Shermer, "Science Is My Savior," accessed Feb. 23, 2007, http://www.science-spirit.org/article_detail.php?article_id5=20. Shermer's article appears to be no longer available online.

xiii. "About the International Darwin Day Foundation," International Darwin Day Foundation website, accessed June 12, 2012, http://darwinday.org/about-us/.

xiv. Quoted in John G. West, *Darwin Day in America: How Our Politics and Culture Have Been Dehumanized in the Name of Science* (Wilmington: ISI Books, 2007), 210.

xv. Comment by "Evolution171," International Darwin Day Foundation website, accessed June 12, 2012, http://darwinday.org/about/#comment-444861134.

xvi. "America's Evolutionary Evangelists," accessed June 12, 2012, http://evolutionaryevangelists. libsyn.com/.

xvii. Michael Dowd, *Thank God for Evolution: How the Marriage of Science and Religion Will Transform Your Life and Our World* (New York: Plume, 2009).

xviii. Description at The Great Story website, accessed June 12, 2012, http://evolutionaryevangelists. libsyn.com/. Emphasis in the original.

xix. See, in particular, the links on the home page for "Great Story Parables," "Great Story Beads." "Songs," "Children's Curricula," and "Group Study Support Materials," The Great Story website, accessed June 12, 2012, http://www.thegreatstory.org/.

xx. Michael Dowd, "Thank God for the New Atheists," *Skeptic Magazine* 16, no. 2 (2011): 29.

xxi. Ibid., 29–30.

xxii. Comments by Loyal Rue, highlighted on The Great Story website, accessed June 12, 2012, http://www.thegreatstory.org/what_is.html. Emphasis added.

xxiii. Phillip Johnson, "In the Beginning Were the Particles," Grace Valley Christian Center, March 5, 2000, accessed June 12, 2012, http://gracevalley.org/sermon_trans/Special_Speakers/In_Beginning_Particles.html.

xxiv. Lewis, *That Hideous Strength*, 79.

xxv. Richard Dawkins, "Is Science a Religion?" *The Humanist* (January/February 1997), accessed June 12, 2012, http://www.thehumanist.org/humanist/articles/dawkins.html.

xxvi. C.S. Lewis, "Christian Apologetics," *God in the Dock*, ed. Walter Hooper (Grand Rapids: Eerdmans, 1970), 95.

xxvii. C.S. Lewis, "God in the Dock," *God in the Dock*, 241.

xxviii. Lewis, *That Hideous Strength*, 99–100.

xxix. C.S. Lewis, *Surprised by Joy* (New York: Harcourt Brace Jovanovich, 1955), 203.

xxx. Entry for Thursday, May 25, 1922, C.S. Lewis, *All My Road Before Me: The Diary of C.S. Lewis, 1922–1927,* ed. Walter Hooper (San Diego: Harcourt Brace Jovanovich, 1991), 41.

xxxi. Ibid., 44.

xxxii. C.S. Lewis, *The Pilgrim's Regress* (New York: Bantam Books, 1981).

xxxiii. Armand M. Nicholi, Jr., *The Question of God: C.S. Lewis and Sigmund Freud Debate God, Love, Sex, and the Meaning of Life* (New York: Free Press, 2002), 5; Kathryn Lindskoog, *Finding the Landlord: A Guidebook to C.S. Lewis's* Pilgrim's Regress (Chicago: Cornerstone Press, 1995), 31–33.

xxxiv. Ibid., 48.

xxxv.   Ibid., 49.

xxxvi.   C.S. Lewis, "Bulverism," *God in the Dock*, 271.

xxxvii.   Ibid., 272.

xxxviii.   Despite Lewis's withering critiques of Freud, he did not reject psychoanalysis out of hand, nor the obvious truth that many of our beliefs may be influenced by non-rational factors. See C.S. Lewis, *Mere Christianity* (New York: Macmillan, 1960), 83–84; C.S. Lewis, *Letters to Malcolm: Chiefly on Prayer* (San Diego: Harcourt Brace Jovanovich, 1964), 34.

xxxix.   Ibid., 64.

xl.   Ibid., 65–66.

xli.   Lewis, "Funeral of a Great Myth," 88.

xlii.   Ibid., 89.

xliii.   West, *Darwin Day in America*, 86-88, 122–62.

xliv.   Julian Huxley, "Eugenics and Society," *Eugenics Review*, 28, no. 1 (1936): 30–31, accessed June 12, 2012, http://www.thegreatstory.org/what_is.html.

xlv.   See Leo Alexander, "Medical Science under Dictatorship," *New England Journal of Medicine* 241, no. 2 (1949): 39–47, accessed June 13, 2012, doi:10.1056/NEJM194907142410201.

xlvi.   For more information about the Piltdown hoax, see Frank Spencer, *Piltdown: A Scientific Forgery* (New York: Oxford University Press, 1990). For Lewis's reaction to the exposure of Piltdown, see the discussion in chapter six of the present book.

xlvii.   West, *Darwin Day in America,* 88–92.

xlviii.   Ibid., 268–90.

xlix.   Lewis, *Surprised by Joy*, 174.

l.   C.S. Lewis, *The Discarded Image* (Cambridge: Cambridge University Press, 1964), 17.

li.   Lewis, *Abolition of Man*, 88.

lii.   J.B.S. Haldane, Daedalus, or, Science and the Future (1923), accessed June 13, 2012, http://www.marxists.org/archive/haldane/works/1920s/daedalus.htm.

liii.   J.B.S. Haldane, "Dialectical Materialism and Modern Science: IV. Negation of the Negation," *Labour Monthly* (October 1941): 430–432, accessed June 12, 2012, http://www.marxists.org/archive/haldane/works/1940s/dialectics04.htm.

liv.   Lewis, *Abolition of Man*, 89.

lv.   Ibid., 83.

lvi.   Ibid., 77.

lvii.   Ibid., 84.

lviii.   For a discussion of Lewis's critique of both communism and fascism, see John G. West, "Communism and Fascism," in Jeffrey D. Schultz and John G. West, *The C.S. Lewis Readers' Encyclopedia* (Grand Rapids: Zondervan, 1998), 126–27.

lix.   Ibid., 85.

lx.   Walter Hooper, *C.S. Lewis: A Companion & Guide* (San Francisco: HarperSanFrancisco, 1996), 231–32.

lxi.   Ironically, the British government in the 1990s actually created a controversial government bureaucracy with the same acronym to ration health care. See the website of the National Institute for Health and Clinical Excellence (NICE), accessed June 12, 2012, http://www.nice.org.uk/. For criticism of the real NICE see "Of NICE and Men," *The Wall Street Journal* (July 7, 2009), accessed June 13, 2012, http://online.wsj.com/article/SB124692973435303415.html.

lxii.   Lewis, *That Hideous Strength*, 42.

lxiii.   Ibid., 102, 69.

lxiv.   Ibid., 87.

lxv.   Ibid., 88.

lxvi.   Ibid., 85–8.

lxvii.   George Sayer, *Jack: A Life of C.S. Lewis* (Wheaton: Crossway, 1995), 304. For J.B.S. Haldane's negative review of *That Hideous Strength*, see Haldane, "Auld Hornie, F.R.S.," *Modern Quarterly* (Autumn 1946), accessed June 14, 2012, http://www.lewisiana.nl/haldane/#Auld_Hornie. For Lewis's reply, see "Reply to Professor Haldane," in *Of Other Worlds: Essays and Stories*, ed. Walter Hooper (New York: Harcourt Brace Jovanovich, 1966), 74–85.

lxviii.   Julian Huxley, UNESCO: Its Purpose and Its Philosophy (Preparatory Commission of the United Nations Educational, Scientific and Cultural Organisation, 1946), 8.

lxix.   C.S. Lewis to Dan Tucker, Dec. 8, 1959, in *The Collected Letters of C.S. Lewis,* ed. Walter Hooper (San Francisco: HarperSanFrancisco, 2007), vol. III, 1104; Lewis, "Is Progress Possible? Willing Slaves of the Welfare State," *God in the Dock,* 311–16.

lxx.   Lewis, "Is Progress Possible? Willing Slaves of the Welfare State," 314.

lxxi.   Ibid., 316.

lxxii.   Ibid., 315.

lxxiii.   Kenneth P. Green and Hiwa Alaghebandian, "Science Turns Authoritarian" (July 27, 2010), accessed June 12, 2012, http://www.american.com/archive/2010/july/science-turns-authoritarian.

lxxiv.   David Klinghoffer, "The Stem Cell War," *National Review Online*, April 13, 2011. accessed June 13, 2012, http://www.nationalreview.com/articles/264551/stem-cell-war-david-klinghoffer#; Wesley J. Smith, "The Wrong Tree," National Review Online, May 13, 2004, accessed June 13, 2012, http://www.discovery.org/a/2039; David A. Prentice, "Current Science of Regenerative Medicine with Stem Cells," *Journal of Investigative Medicine* 54, no. 1 (January 2006): 33–37.

lxxv.   Lee Silver, *Remaking Eden* (New York: Harper Perennial, 1998).

lxxvi.   Ibid., 13.

lxxvii.   Ibid., 277.

lxxviii.   Quoted in Kyle Drennen, "NBC: It's 'Pro-Science' to Abort Children with Genetic Defects," June 12, 2012, accessed June 12, 2012, http://www.lifenews.com/2012/06/12/nbc-its-pro-science-to-abort-children-with-genetic-defects/.

lxxix.   Quoted in Leo Hickman, "James Lovelock: Humans are too stupid to prevent climate change," March 29, 2010, accessed June 12, 2012, http://www.guardian.co.uk/science/2010/mar/29/james-lovelock-climate-change.

lxxx.   Eric R. Pianka, "The Vanishing Book of Life on Earth," accessed June 13, 2012, http://www.zo. utexas.edu/courses/bio373/Vanishing.Book.pdf; Jamie Mobley, "Doomsday: UT prof says death is imminent," *Seguin Gazette-Enterprise*, Feb. 27, 2010.

lxxxi.   Daniel Lieberman, "Evolution's Sweet Tooth," *The New York Times*, June 5, 2012, accessed June 13, 2012, http://www.nytimes.com/2012/06/06/opinion/evolutions-sweet-tooth.html.

lxxxii.   Robert Pear, "Obama Reaffirms Insurers Must Cover Contraception," The New York Times, Jan. 20, 2012, accessed June 12, 2012, http://www.nytimes.com/2012/01/21/health/policy/ administration-rules-insurers-must-cover-contraceptives.html.

lxxxiii.   Lewis, *Magician's Nephew*, 9, 11, 19.

lxxxiv.   Ibid., 44–55.

lxxxv.   Ibid., 14–149.

lxxxvi.   Ibid., 159–160.

lxxxvii.   Chapter 1 of *The Magician's Nephew* says it was set when "the Bastables were looking for treasure in Lewisham Road." Ibid., 1. The Bastables were characters in stories published by Edith Nesbit starting in 1899.

lxxxviii.   Lewis, *Abolition of Man*, 90.

lxxxix.   Lewis, *The Problem of Pain* (New York: Macmillan, 1962), 134.

xc.   Lewis, *Abolition of Man*, 89–90.

xci.   See discussion in West, *Darwin Day in America*, 373–74.

xcii.   Ibid., 370–373. Also see Michael Behe, *Darwin's Black Box: The Biochemical Challenge to Evolution,* 2nd ed. (New York: Free Press, 2006); Stephen Meyer, *Signature in the Cell: DNA and the Evidence for Intelligent Design* (New York: HarperOne, 2009); Bruce L. Gordon and William A. Dembski, eds., *The Nature of Nature* (Wilmington: ISI Books, 2011).

xciii.   West, *Darwin Day in America*, 274–375; Mario Beauregard and Denyse O'Leary, *The Spiritual Brain* (New York: HarperOne, 2007).

xciv.   George Gilder, "The Materialist Superstition," *The Intercollegiate Review* 31, no. 2 (Spring 1996): 6–14.

xcv.   Lewis, *Abolition of Man*, 87.

Resources for Naked Eye Astronomy

Suggestions for help getting started with naked-eye astronomy:

1. *Star Gazers*: A short, weekly video program designed to help amateurs with naked-eye astronomy. Browse through 3-4 past episodes for help with everything from phases of the Moon to eclipses to seasonal constellations. Go to http://stargazersonline.org/index.html.

2. *What's Up?*: A monthly program found on the NASA/Jet Propulsion Laboratory website. There are many other worthwhile tools on this site, but this one is there to help you with naked-eye viewing. Go to http://www.jpl.nasa.gov/, scroll down to "What's Happening Now," and choose "videos," or just search "jpl what's up."

3. *Sky & Telescope*: This website has many helpful features, including "This Week's Sky at a Glance." Stick to the features that apply specifically to naked-eye astronomy, such as . . . https://skyandtelescope.org/observing/sky-at-a-glance/.

4. *Astronomy.com*: has a feature called "The Sky This Week." Limit your use to naked-eye observing and not their hints for using binoculars or telescopes. Go to https://astronomy.com/observing/sky-this-week.

5. *Stardate.org*: like several other sites, this one includes a "night sky" page to help point out important objects or events for a given week or month. Go to https://stardate.org/nightsky.

6. *Stellarium*: This is a free simulation program that allows you to see what to expect in the night sky from any location and any time. Go to https://stellarium.org/ and download the program.

7. *Sky Map Online*: This website hosts an online planetarium program. See the sky from anywhere in the world at any time of day. Go to http://www.skymaponline.net/default.aspx.

8. *Sky Maps*: Each month, you can print out a free copy of these "maps," with versions that correspond with any location on Earth. So while most of us would print out the "Northern Hemisphere" sky map, there are versions for equatorial and southern hemisphere viewers as well. Go to http://www.skymaps.com/index.html.

9. *Space.com*: another versatile website with lots of good content. Scroll through and choose the features that apply to naked-eye observations, such as maps of the night sky and summaries of the best objects to look for in a given month. To get started, go to https://www.space.com/skywatching.

10. *In-the-sky.org*: this site has many features, so be sure to limit yourself to the tools that apply to naked eye astronomy. Go to https://in-the-sky.org/.

Tables of the Brightest & Nearest Stars

Table C.1 The Twenty Brightest Stars

Star	Apparent Magnitude	Distance (light-years)	Absolute Magnitude	Spectral Type
Sirius	−1.44	8.6	1.5	A1
Canopus	−0.62	310	−5.4	F0
Arcturus	−0.05	37	−0.6	K2
Rigel Kentaurus	−0.01	4	4.2	G2
Vega	0.03	25	0.6	A0
Capella	0.08	42	−0.8	G8
Rigel	0.18	800	−6.6	B8
Procyon A	0.40	11.4	2.7	F5
Achernar	0.45	144	−2.9	B3
Betelgeuse	0.45	520	−5.0	M2
Hadar	0.58	500	−5.5	B1
Altair	0.76	17	2.1	A7
Aldebaran	0.87	65	−0.8	K5
Spica	0.98	260	−3.6	B1
Antares	1.06	600	−5.8	M1
Pollux	1.16	34	1.1	K0
Fomalhaut	1.17	25	1.6	A3
Deneb	1.25	1500	−7.5	A2
Acrux	1.25	320	−4.0	B1
Mimosa	1.25	352	−4.0	B1

Table C.2 The Twenty Nearest Stars				
Star	**Distance (light-years)**	**Apparent Magnitude**	**Absolute Magnitude**	**Spectral Type**
Proxima	4.2	11	15.5	M6
a Centauri A	4.4	−0.01	4.3	G2
a Centauri B	4.4	1.35	5.7	K0
Barnard's Star	5.9	9.5	13.2	M4
Wolf 359	7.8	13.5	16.6	M6
Lalande 21185	8.3	7.5	10.5	M2
Sirius A	8.6	−1.44	1.5	A1
Sirius B	8.6	8.44	11.3	A0 (WD)
Luyten 726 A	8.7	12.6	15.4	M6
Luyten 726 B	8.7	13	15.8	M6
Ross 154	9.7	10.1	13.3	M3
Ross 248	10.3	12.3	14.8	M4
e Eridani	10.5	3.7	6.2	K2
Lacaile 9352	10.7	7.35	9.8	M2
Ross 128	10.9	11.12	13.5	M4
Luyten 789	11.2	12.52	14.6	M5
61 Cygni A	11.4	5.2	7.5	K5
Procyon A	11.4	0.4	2.7	F5
Procyon B	11.4	10.7	13	A0 (WD)
61 Cygni B	11.4	6.1	8.3	K7

Creation Sites & Resources

Creation Resources:

1. Answers In Genesis https://answersingenesis.org/

2. The Institute for Creation Research https://www.icr.org/

3. Awesome Science Media https://www.awesomesciencemedia.com/

4. Biblical Science Institute https://biblicalscienceinstitute.com/

5. Creation Astronomy https://www.creationastronomy.com/

6. Is Genesis History? https://isgenesishistory.com/

7. Creation Ministries International https://creation.com/

8. Creation Research Society https://www.creationresearch.org/

Video Credits

Chapter 1

Contributed by Charley Dewberry. © Kendall Hunt Publishing Company

Chapter 3

Contributed by Mark McGinniss. © Kendall Hunt Publishing Company

Chapter 4

Contributed by Jakob Sauppe. © Kendall Hunt Publishing Company

Chapter 5

Copyright © Leo (Jake) Hebert, III. Reprinted by permission.

Chapter 6

Copyright © Leo (Jake) Hebert, III. Reprinted by permission.

Chapter 7

Copyright © Leo (Jake) Hebert, III. Reprinted by permission.

Chapter 8

Contributed by Marcus Ross. © Kendall Hunt Publishing Company

Contributed by Timothy R. Brophy. © Kendall Hunt Publishing Company